ECONERGY:

THE CONCEPT OF PLENTY

by

Robert D. Bowers

Published by
R & E RESEARCH ASSOCIATES, INC.
Publishers
936 Industrial Avenue
Palo Alto, California 94303

Library of Congress Card Catalog Number
82-80466

I.S.B.N.
0-88247-655-6 (Hardcover)
0-88247-667-X (Softcover)

cover design by
Valeries Zwinscher

ECONERGY — THE CONCEPT OF PLENTY

Econergy is a term applied to businesses that produce from that great storehouse of energy and products that are renewable. It should not be considered a term of the technology but rather one of application by the business entity.

Econergy is the concept of using renewable resources for the production of food, clothing, shelter and energy. It is the needed concept between today's technology and the businesses that must be developed NOW.

By changing our minds and our dependencies on the fossil fuel economy, we can change from the concept of less and less to ECONERGY — THE CONCEPT OF PLENTY.

DEDICATION

This concept is dedicated to just one person called "you". It was written from the point of view that just you and I sit down to discuss the world's greatest crisis. Therefore, I write in the first person because that is how we would discuss this problem and solution in such a situation.

If there is a group to which I would make a dedication, it is the group of men and women who think of themselves as "small business people."

With this dedication there is a responsibility. Those who become ECONERGISTS are charged with expanding the concept and with generating what can become the new era of world prosperity.

ACKNOWLEDGEMENT

There are so many to whom I owe so much. Those who aimed my thinking toward renewable resources about thirty-five years ago, those who helped me succeed, those who helped me fail; ALL contributed to my concept of ECONERGY and to this book.

There are also those helpful people in universities, industry and other organizations in the states of Minnesota, Nebraska, Wyoming, Colorado, Utah, Nevada, California and Arizona in which I traveled during the writing of this book.

There are my friends who endured much as I tested ideas on them over the years.

And like almost every man who writes a book, I have a wife and family who helped and gave heavily toward the project.

TABLE OF CONTENTS

PREFACE

PREMISE OF BOOK

PREMISE 1 — CRISIS

An economic, political and social crisis is rapidly building on a world scale. Its magnitude is greater than all of the combined historical disasters of war, depression and famine. It is fueled by the total dependence of the producing world upon a finite supply of fossil fuel and upon the few entities that control that supply.

PREMISE 2 — CONSERVATION

All surely agree that conservation is good because it results in preservation. However, when it is used in the sense of less and less, the result on an economy is destructive. With a change of emphasis to conservation of political, economic and social institutions, and with the understanding that these institutions are built upon a foundation of fossil fuel, it seems logical to change our dependencies from fossil fuel to those resources that are not depleted when used. This is a concept of affluence and sufficiency. It is "Econergy — The Concept of Plenty."

PREMISE 3 — TECHNOLOGY

Many wait for the great technological breakthrough that is to save the world from disaster. The fossil fuel economy is too technical, and its deposits of raw material are too unevenly divided on earth, for two-thirds of the world population to use.

The result is reflected in economic misery. The great break-through has already been made, not in higher technology but in the low to medium technology already available to utilize renew-able resources. A lower level in technology allows greater num-bers of people to participate.

PREMISE 4 — THE SMALL BUSINESS

The small business community has never before been called upon the save the world from disaster. This sector has never been considered significant when compared to giant-size government and giant-size industry. However, never before has there been an economic need that fits the giant institutions so badly. The renewable resources are available everywhere on earth. As such, it is almost impossible to control the resource. Therefore, small industry of local or regional size can compete successfully with world-size companies. The emphasis should change from the manufacture and sale of energy to that of the manufacture and sale of energy-producing equipment to make the consumer energy self-sufficient. The breakthrough, then, is to be in the development of a new level of small business rather than in science and technology. (This does not indicate less scientific research but more research on the simpler technologies for the small business entity.)

PREMISE 5 — CONCEPT

A book on concept leaves a great deal to be added by the reader. This is a way of saying that such a book is directed to those with ideas of their own. It is this group of readers that can apply the concept to what they believe, to what they can do, and to what will be profitable to them. For me to make a step-by-step procedure would be for me to direct the reader to the things that are important to me. Almost everyone has strong ideas of his own. All can let their great beliefs and desires soar as they read about econergy.

PREMISE 6 — A GREAT NEW LEVEL OF INDUSTRY

Many should become involved in this new level of industry

to convert energy, chemicals and products from renewable sources. Producing in these ways will give abundance in food, clothing, shelter and energy. In so doing, the economic impact of this new level of activity will create a higher affluence for all people.

The following pages discuss what to do, who should do it, when to do it, how to do it, and where to do it. "If" it should be done is left up to you.

PREMISE 7 — FOOD, CLOTHING, SHELTER AND ENERGY

The four basic needs of mankind may seem mundane to many of a modern industrial nation. However, all endeavor is aimed toward attainment, which in turn, results in a total economy. The quantity and degree of refinement that a society obtains of food, clothing, shelter and energy, together with services and frivolities, is described by the relative terms of low, medium or high standard of living. The total economy, or gross national product, is generated by *producing* and *using* the fruits of the search for these basics. Therefore, *more* is better than *less*, since more generates greater wealth for all. If *more* is to be maintained we must produce from the renewable storehouse.

Overuse of the four basic terms in this book may seem redundant but these needs are the track on which econergy will move through its many phases and it is good to keep our minds on the direction that we must travel.

INTRODUCTION

This book is about how you and I, as just two ordinary people, can be fulfilled in our physical needs, can build a good economy, can maintain a happy culture for our families and loved ones, and can become proud and important members of our community; not because we did those things for just ourselves but because by doing so we helped our neighbor, our country and, more importantly, because we helped those people around the world who are now so miserable from need for food, shelter, clothing and energy.

The book is about you and me and how we can win in this next industrial and social revolution into which the world is now submerging.

The book is about how we can keep some of the creature comforts to which we have become so attached. It is about how we can get the energy we need to keep our personal transportation. It is about how we can continue our prized mobility and our affluent lives. It is about how this can be done by not taking away the same chance for affluence from the poorer nations and their hungry people.

The book is about how you and I can become free and independent from the great dependence we now have in energy, food, and other raw material on super powerful, international corporate structures and on super powerful, foreign control of energy.

It is about how you and I can do something about our plight regardless of inflation, depression or political power.

As you read this I hope you will be struck with its idealism. This value is usually looked upon as being naive and silly. I present this form of idealism as the ingredient missing in today's world of computers and numbers. Idealism gives us something to believe in and it is the thing that will make tomorrow more liveable. It is the thing that is far ahead of the categorizations, of the symbols and numbers into which we are being pushed as persons so that the computer can be right.

Idealism aside, the cold hard facts are that you and I are on the threshold of being denied the energy we need to maintain our culture and our economy. One country, one company, one man or one politician can stop the flow of fossil fuel any hour. An emotion, a war, a bid for world power could trigger such a disaster. You and I have no control of this, but we could have control if we produced new source energy for ourselves from simple equipment which needs no new invention to be used right now. Small companies can produce this equipment.

I am just a part of the people of this world who can best be described as "just folks." I don't know everything about the world of renewable raw material but I have learned some things about it over the years. I have not written a complete report here on the subject either. I will wait for some of you to write the more detailed presentations as the concept develops and matures.

I pledge what I know and believe in to help you and your friends, and I urge you to help those who work at building the new industries described in this book.

You and I can fill a gap of importance now. We can be ECONERGISTS for just the price of saying that we are one, and for believing in *Econergy — The Concept of Plenty*.

At this point, you must know that ECONERGY has nothing to do with a new science. It is a new "business" concept and involves the utilization of all the sciences and the "old" arts as they relate to the sun. So you ask, "Why don't you call your concept something like 'Solar Business'?"

This would be a good name except for the fact that most of us think of solar energy as relating to sunbeams that can only be used when the sun shines. The real fact is that solar energy applies to just about everything that we now use, including the fossil fuels (petroleum and coal). The sun stores its energy in many forms — plants, animals (including man), rain, wind, heat, coal, oil, rivers, tides and currents. If we assume that the earth, together with its minerals, was formed and developed by the forces of our sun, and perhaps other suns, we will come to the conclusion that just about everything is SOLAR ENERGY. So, I refer to the total complex of such raw materials as *RENEWABLE RESOURCES* instead of *Solar Energy*.

The fossil fuels (made from ancient organics by geological processes in time) and the minerals of earth are finite. Once used they are gone forever. It is this group of raw materials that generate today's economy. It is, also, this group of basics that must be replaced by renewable-type resources over the next generation.

It is in the area of renewable resources that ECONERGY dwells. As the name implies, it is the economy (econ) becoming involved with the many new energy forms and services (ergy). This indicates that *business* get involved instead of leaving it to government and technical research community. In other words, it is the concept of taking today's level of simple technology and producing now.

A definition of the concept:

> ECONERGY is the concept of using renewable resources for the production of food, clothing, shelter and energy. It is the business link between today's technology and the new market that must be developed.

The method of accomplishing this:

> By changing our minds and our dependencies on the fossil fuel economy, we can change from the concept of less and less to ECONERGY — The Concept of Plenty.

There are three general areas of the heliosciences (sun sciences):

(1) *HELIOCHEMICAL* — This is the big area of photosynthesis which produces and maintains all plant and

animal life (organic). This form of sun energy is one of the oldest to be used by man. The total spectrum in this area is broad and is ECONERGISTIC with the exception of the fossil fuels which also fall in this category.

(2) *HELIOELECTRIC* — There may be many ways to use this property of the sun, but one of the current ways is through the photovoltaic cell (solar cell). Electric current flows when radiation from the sun dislodges electrons from a properly treated silicon cell.

(3) *HELIOTHERMAL* — For the use of "low grade" energy, this property of the sun has the most widespread application with the lowest technical requirements. The sunshine, when it strikes a dark surface, is absorbed and can be stored as heat by a solid, liquid or gas. This is the property that will heat and cool the home, store and factory in the near future.

Therefore, prepare to think a little differently. Believe that simple technology is possible. Believe that small industry is possible. Think like tomorrow's business person must think. Think ECONERGISTICALLY.

Robert D. Bowers

PART ONE

CONCEPT

TO UNDERSTAND THE BASIS

*Know all the concept that you can
because it will lead your mind to tomorrow's new activity.*

CHAPTER I

ECONERGY — THE CONCEPT OF PLENTY

It has taken thousands of years to build the so-called affluent society. To destroy this state of well-being or to let it dissolve into the nothingness of world misery is not the answer. The thrust should be toward higher living standards and simple, understandable means of obtaining it for everyone.

Econergy is the business activity of converting renewable resources to the basic needs of mankind. The roots lie in the minds of the common man and woman, and in the understanding and support that they give to their local entities that enter this business.

Econergy uses the most economic raw materials. They are those that are renewable. The sun, the wind, gravity, water, thermal gradients, hydrogen, organic material, muscle power and all of agriculture are a few obvious sources of the renewable. These resources can be used continuously for thousands upon thousands of years by billions and billions of people for their food, clothing, shelter and energy, and their use will maintain an economy to support their culture. Unlike fossil fuel and mined material, they cannot be depleted. With proper husbandry they will last as long as the earth is a proper habitat for people.

How often in recent years have we heard from government, industry and the academic community that we will have to

3

reduce our standard of living and exist on less and less. Nearly all the ideas on the subject of energy and raw materials are negative. Everyone looks to their government for solutions. The thrust has been toward laws which reduce production and consumption. We are told of many things "not to do." Seldom have we heard of things "to do."

Government has become convinced that adjustments in tax rates and tax methods will solve the problems of inflation, depression, production, energy and monetary crisis. The so-called energy bills in the nation today are directed toward taxes and price controls as though they had something to do with the existence of usable energy.

The benefit is to those major money pools now in the fossil fuel business. Profits in the oil and coal industry are great and the companies in this business, understandably, want to keep it that way. Their great economic power and their effective Washington lobby have good control at the administrative, legislative and judicial levels of government. The result is the maintenance of the status quo which is a way of saying that you and I are captive and dependent upon these few companies for our livelihood and in fact our very lives.

This is not to say that our lives have not been great as a result of this system, because they have. We have, in this Western World, lived better with more comfort and wealth than could have been imagined a generation ago. Letting ourselves be captured into such affluence has been wonderful.

Now we must face a real shocking truth. The fact is that we have been having a real ball by drawing on what was a savings account (oil reserves) that cannot be replenished by a new deposit. We now also realize that the account we are drawing heavily on is not really our own but that of a world that may not always be a "friendly banker."

This present system keeps you and me dependent, but those in the poorer nations are even more so. The basis of this is not only controlled energy but its effect on food production. The production, processing, distribution and preparation of food consumes large quantities of energy. It has been estimated that it takes one-half calorie of petroleum for each calorie of food consumed in the Western World. It seems a major error in judgment to tie such an important renewable resource such as food to such a finite resource as fossil fuel.

Econergy is a concept that can change these dependencies. It is for the rich and poor nation, the rich and poor person, the communist nation and the democratic nation. It takes nothing from anyone, but by sharing the concept gives freedom from dependency to everyone. Production of the things that people need can be done by individuals and local or regional business entities.

The present multinational giants must continue doing what their large capital formations do best, but local econergistic-type energy capability will keep prices fair as scarcity forces fossil fuel prices up. Change to the new system will take a generation. This challenge to the status quo will not require civil disobedience, public demonstration or war. It will be gradual, orderly, productive and profitable.

Change is coming regardless of the form it takes. War is the probable tool now being considered to force this change. People will riot and nations will go to war to capture deposits and production of fossil fuel. Poor nations will go to war to become rich. Rich nations will go to war to protect their wealth and economy. The dependency of one on another in a world of self-centered nationalism is a short-fused bomb. Equal availability of food and energy will defuse this bomb.

By changing our minds and our dependencies on the fossil fuel economy, we can change from the concept of less and less to ECONERGY — the concept of plenty. A million small business entities doing a million small things in new source energy will solve the energy crisis and produce abundance.

Government and industry tell us that answers to the energy problem will come by the year 2000 A.D. By that time, we will have used most of the available cheap petroleum and will be converting to another fossil fuel — coal. True, there is a lot of coal. But it, too, is finite. Like petroleum, when we use a pound, it is gone forever. It is a capital asset, not an income item. We should use it carefully in our energy and chemical industry so that it might last for thousands of years instead of a couple of lifetimes.

Therefore, everything in this book is aimed at helping you become ECONERGY oriented. There is no attempt to write a technical report on economics, business or the sciences. The attempt is to point out how simple the answers are when we decide to solve just your problem and my problem, and how difficult it is to solve the national or world problem. When enough individuals and communities solve their own problems, each of them, one at a time, there will be no national or world problem.

CHAPTER II

ECONERGY — THE ECONOMIC IMPACT

Since the beginning of the industrial revolution, there has never been such a great chance for the development of small business by small business people. There has never been a development in any producing industry that fits so many geographic areas or that promises such good social results. There has never been a new basic industry that will allow as much immediate participation by so many new and relatively untrained people.

When we think of positive economic impact, we usually think of the added economy to a community or a nation that resulted from the addition of a new industry. The new jobs, the new products, the new services and the multiplier effect of the new industry have been most desirable. Cities have worked hard to have these industries locate in their area. The result was economic growth.

During this period of growth in the city, the rural area grew very little and, in many cases, it declined. Fewer and fewer were required to operate the agricultural sector. As jobs declined in the rural community, the rural worker moved to the new job prospect in the city.

Crime, social problems, pollution, energy shortage, transportation problems, high taxes, municipal management breakdowns and welfare problems are a few co-results of the good economic impact. Most of these problems are the result of big-ness — bigness of the industrial sector, bigness of the city, bigness in the labor pool, bigness in the service industries. The city directs the commerce of the rural community so the problems of the city become those of all. Times of depressed economy make the turmoil greater and point out the problems even more.

At this point, I must make a statement about the big multinational companies that are mentioned so often. Many would wish their demise. I do not see their decline nor do I want to see it. There is a place for large money pools and for the things that they can do. It takes a large company to supply cars, trucks, trains, airplanes, petroleum and world transport facilities. The point is that they are not needed to supply *all* food and *all* energy of the future. At best, these big money pools are a type of internal bank with executive and production management. If these companies are forced to become less monopolistic because of the availability of renewable energy and other renewable raw materials, they could become ideal financial and management companies to support the development of small industry around the world. They do not need absolute control of raw material, production, labor and marketing to have a profitable business. The investment and lending of money and the sale of manage-ment and production technology could prove rewarding in the future.

So, let's not legislate against the large companies. Let's not suggest, for example, that gasoline would be cheaper if the oil companies had to spin off their processing and marketing divisions. This would surely raise the price of consumer products.

The new econergy industries will be relatively small at first. Some will grow and become large in the future. But never will a few control *all* the production, because the raw material (sun), or sun products, are fairly divided to all areas. The dis-persion of industry and jobs resulting from this fact will benefit all regions.

The industry will naturally develop in the rural sector for the simple reason that sun, wind, water and agriculture are more available away from the major cities. But cities must not decline

8

in importance either. There must be a place for financial centers, labor pools, distribution centers, transportation, management bases, services and cultural outlets.

So what is the economic impact of the new econergy industries?

It is the fact that it will force gradual economic change without great shock waves. Every life and every business today is tied to fossil fuel. The economy could not survive a political move to change from established fuel sources rapidly.

Small private business and the profit incentive, in competition with fossil fuel producers, can build a healthy new group of industries that can stand on their own merit.

The greatest help from all levels of government to support this needed change would be to withdraw the exclusive license to do business, to cancel regulations which promote conformity, and to simplify security regulations (regulations for sale of business interests such as stock) which now cause filing costs to soar beyond reason for new, small companies.

The insertion of a new level of industry in the economy will generate new wealth, new jobs and new confidence in private enterprise. People can be busy and interested in the new hope around the world.

The city, the rural area, the rich nation and the poor nation, all can participate on the basis of availability of energy and food.

It will reduce the need for war to capture fossil fuels.

It will spur new technology which in turn will spur more new wealth.

The greatest result on an economy based on econergy would be its social impact. Poor and hungry nations can use this concept and compete on the basis of equal availability of raw material with richer nations. All have great resources in the renewables.

CHAPTER III

NEW INDUSTRY PROSPECTS

The industries in the new field of econergy are simple and inexpensive to build if they are kept small and local. Massive, monopoly-sized operations can only retard the realization of this concept.

This chapter deals with some of the business possibilities for new econergistic enterprise. Some are ready for commercial application now. Some require a pilot plant or testing stage prior to full production. Few of them require anything to be invented or discovered. Some of them are applied best to small new business entities. All of them would be a good expansion for existing small to medium sized business. Most of the equipment and processes will, of necessity, be of first generation technique. All can be technically improved as production proceeds.

It has always been difficult to launch a new idea into a going commercial operation. Reasons for this are many but the main problem is that so many individuals are employee-oriented and have dependent-oriented minds. How often have you heard, "If it is so good, why isn't somebody doing it?" or "If it is so good, why are you asking me to get in on it?" or "If it is so good, why aren't the major companies doing it?"

This thinking is in error. Large enterprise cannot be concerned with small beginnings. Their overhead charges require that new product manufacture must have large immediate markets and high profit margins. Small companies can best develop the new industries of the type that we are discussing. As you will note, few of them require a super high level of technology. Therefore, most areas of the world can participate with a little outside help.

Would there be a Ford Motor Company if Henry Ford had waited for the development of the Model "A" car? Of course not, and there will not be new energy sources and new products from renewable resources unless we produce at a Model "T" technology for awhile. Everything has a beginning and an ending. Beginnings of most everything are as embryos. If they have a reason to exist, they will grow and mature. When their time is served, they will die.

A method for making these new embryos exist is the subject of the following chapter. This chapter is concerned with some new areas in which small companies could develop *now*.

Since this book is not designed to be technical, the following overview is only a partial look at a long list of possibilities. The technical descriptions, if any, are meant to be superficial. More technical information is available from the local science teacher, the library, the university.

There are no secrets in the way of the development of a new business if action is applied by an imaginative mind. The application of good ideas with sound business practice will assure success in this new field.

SOLAR POWER

It is usually best to discuss the sun first when considering renewable resources. When all energy is reduced to its real source, the sun is either the basic source or the example. If we ever fully understand the sun and its function, we will have abundance in all the needs of mankind.

All organic material; rivers, tides, currents, thermal gradients, wind, rain, et cetera, are a few of the evidences in which the sun naturally stores its energy on the earth. Therefore, even though it does not shine on one spot all the time, the sun leaves a storehouse of its energy in many forms to be used when needed.

Plant and animal life is one of the greatest usable sources of stored solar energy. This flora and fauna are continuously producing and reproducing. They can be converted mechanically and chemically to about every product used. This area holds much promise in the development of new energy and new products.

"Energy farming" could become a new common term to define those farmers who produce crops for use as energy and chemicals. New crops with new multiple uses are possible. Basic technology is available now to convert the non-food portion of crops (hulls and stalks, etc.) to usable energy. Several applications of this type are discussed later in this section under Crops and Crop Residues.

The *solar furnace* is probably one of the simplest applications of the sun. Almost anyone could build one out of junk, an old storage tank, glass, wood, pipe, a small fan or pump; all could be obtained from the junkyard. A little muscle power will assemble it. Plans are available at most schools and in many magazines and books.

More refined units are being manufactured and are available. In general, they all consist of panes that collect the sun's radiation and store it as hot water or as heated rocks. The heat is carried from the panel by a liquid or by air. The stored heat is then used to heat or cool a building.

For those who have visited one of the solar homes that have been built around the country, they are always impressed with the utter simplicity of the installation. This feature will surely entice many new people into the field with new ideas on product and better marketing concepts. Here is a market that is ready and that exists wherever there is a building. This one does not need to wait. Many new businesses could be started and many are operating *now*. The first big use for solar radiation has been in heating domestic water. This is now a big business and has proven the fact that the sun is a valuable ally in solving the

12

energy crisis.

Solar cells are developed and convert solar energy directly into electric current. They are in use in the space program and are the principal power source of the orbiting satellites. They supply power for years and are perhaps one of the most important recent developments. Costs are still high but new improvements are being made to reduce cost.

Many new industries will emerge using the solar cell. Storage of the energy in batteries, flywheels and as hydrogen gas make possible a wide range in which to work. The feature of energy in the future is the diversity of sources and uses, not the one fossil fuel source of the past.

Most interesting about solar applications is the fact that "economy of scale" does not apply in its use. This means that the home and factory will be self-contained and not connected to public power. The only economy of scale would be employed in the manufacturing of the solar units. THIS IS A FEATURE WITH HIGH ECONOMIC IMPACT AND IN TIME WILL CHANGE THE CONCEPT OF SELLING "POWER" TO ONE OF SELLING "POWER GENERATING EQUIPMENT."

WIND

Wind has been a source of cheap power for thousands of years. *Sailing ships* use it to great advantage. There is a big industry today in making small sailboats, so the art is not dead. However, with scarce and high cost oil, it may be the right time to look at sails on cargo ships again.

There are ways to automate the operation of the sails and there is much known about new ship design and airfoils. There may be good application in the use of river and seagoing barges where a small tugboat may make good time with very large loads.

Those in seashore communities should look into this. There are many old shipyards that need a new concept of an old idea. The idea of keeping transportation at low cost is most attractive. Nothing could do more to increase business than to be able to move goods cheaply. We have been blessed so long with low cost petroleum that we cannot imagine the consequences as the price escalates.

Electric wind generators have been turned for years by propellers. The current is stored in batteries and used as needed. The most practical use of this type of system is in small units for home or factor. There have been many ideas for using wind power on a large power plant scale. I doubt that it will ever be practical for that purpose. However, with improvement, it is practical for special uses and it deserves more thought. There is a place for the production of such units. As other storage is developed in such things as flywheels and hydrogen gas, the demand for efficient units will increase.

The wind could be used to power another type of stored power. That is potential energy using *gravity*. I call this using *weights, hills* and *holes*.

The weight of water rushing downhill through a power generator is the basis for today's hydroelectric system. This was our first major power from solar radiation. The same principle can use wind to keep a supply of water at a high point, to be released and continue to generate power when the wind stops.

A friend of mine in Colorado pointed out the value of this a few years ago. A dry oil well had just been drilled on his ranch. In order to use this expensive hole in the ground, he, with tongue in cheek, proposed to put a wind generator over the hole. As the wind blew, he would generate electricity to pump water for irrigation. At the same time, a weight on a cable would be wound up to the top of the hole. When the wind stopped, the weight would be triggered so that it would start to descend to the bottom of the mile deep hole. On the way down, it would continue to turn the generator. Neither of us worried about the economics of such a system but I thought it was an ideal example of using weights, hills and holes.

Wind is an important force and warrants serious thought for use by small units. It probably will never become a major factor in public power distribution, but a small industry could have a great future with the proper application.

WATER

Water (H_2O) as a chemical holds the most promise as the source of burnable gas. Its two parts of *hydrogen* make it one of the most accessible energy sources. When we consider that two-thirds of the oceans, lakes and rivers are made up of hydrogen, the supply seems limitless.

The features of this fuel are such that it fits well into the econergy concept. When it burns, it turns to water vapor, falls back to the earth as rain to be used again and again and is clean and non-olluting. There is no other burnable fuel so clean and so abundant. This prospect alone should instill optimism in the minds of those who only see gloom on the energy horizon.

As one discusses hydrogen gas as a fuel, a common distrust as to its safety is evident. What about the Hindenburg disaster? Didn't the Hindenburg burn because it was filled with lighter-than-air hydrogen? The answer is yes. But it is not a gas to use in this manner. It didn't cause any more trouble than gasoline would have caused if it could have replaced the hydrogen. We have all learned to live with gasoline even though it is highly explosive. We can learn to live with hydrogen in the same way. With proper handling and proper education, we can expect to use it just as well as we do any other fuel.

The most common way to separate the hydrogen and oxygen of water is by electrolysis. This process has been known for many years and most high school students have seen the process demonstrated in their science classes. We no doubt would have been using this gas for many applications were it not for the fact that petroleum fuel has been so cheap. It is also obvious that the market value of hydrogen would have to be more than the cost of electricity used to produce it. Electricity from solar cells or wind or other heliothermal sources is surely this cheap electric supply.

There are many plans to use it in the future. Most apply the large-scale power plant idea. It has been proposed that hundreds of windmills be placed in the ocean to generate power to make hydrogen, to then pump the hydrogen onshore for use.

Others suggest large atomic plants in the ocean. The cost of such units would be many billions of dollars. Again, we come back to the concept of small personal size or community size electrolysis plants. This size we can handle. And why go to the expense of transporting the hydrogen to market when the raw material, water, is already available wherever people live?

This gas lends itself well as a way to store intermittent energy sources such as the sun and wind. Electricity from solar cells or wind generators could make hydrogen when the sun shines or when the wind blows. The gas would be available when needed.

For a small company who would like to enter a worldwide business, this product offers much. But don't try to get in the business of selling hydrogen gas. Enter the business of making and selling the device to make the gas. Once this is done, the spin-off industries making storage tanks, conversion kits for engines and heating units and many, many new uses will set into motion a whole new world economic order. The emphasis again must be on the small home or local unit —not on the big power plant or refinery.

Once a person gets his mind on hydrogen, it is hard to look at some of the other energy sources created by water. However, there are some good ones, and since the energy supply of the future must be diverse, we should use them all where applicable.

Geothermal power has been used since 1904 in Italy. Other countries have electric power plants built around the hot water and steam of underground reservoirs. This may be a science that will develop to a point that drilling can be done to the molten rocks and gases deep in the earth. Costs of geological exploration and development may be high but it should be done.

Thermal Gradients is the term used to describe the temperature variation between warm ocean surface water and the cool ocean depths. If the warm surface water is pumped through a heat exchanger, it will cause a low boiling point liquid to boil. Ammonia is such a liquid. Its vapor, when run through a turbine, would power an electric generator. The deep ocean water would then be used to condense the ammonia back to a liquid for reuse

16

in the cycle.

This system is applicable to units of various size, but a small unit would be most important to an underdeveloped community in the tropics. Uses such as this will make the present "have nots" very independent and give them local power for new industry that they need so badly. Again, here is a future business for those who would build small home and community units. Sell the units, not the electricity.

Rivers, tides and currents is a fourth area in which water works well in the generation of usable energy. The principle is so well known that I will not attempt to describe it here. We have all been exposed to the hydroelectric power plant, and tides and currents can be harnessed in much the same way.

It is my opinion that these systems are only applicable to large power combines. With the prospect for so many small personalized units that will make the home and factory a self-contained energy unit, I doubt the need for adding many more large power monopolies to the world system.

CROPS AND CROP RESIDUES

Crops and crop residues are the domain of bio-mass or chemurgy. The concept has been around for many years but has not had widespread understanding. Bio-mass and chemurgy deal with the conversion of renewable organic material to usable energy and chemical products. The concept is usable by econergists, except that the commitment is to organics only.

There have been chemurgic (or bio-mass) industries long before the terms were used. All of agriculture is chemurgic. An example — the paper industry converts cellulose (trees) to paper. A livestock rancher is in the same business. He converts cellulose (grass) to protein food through the ruminant animal (cow, sheep or goat). Most important is the fact that agriculture is really storing and converting solar energy in its most natural form. All of the petroleum and coal was made from the things that grow. Plant life is one of the greatest natural usable storehouses of solar energy.

This writer was introduced to the chemurgic concept soon after World War II. Dr. Michael Pijon was doing research for the Chemical Foundation of New York. He passed on the vision of a great new chemical and energy business based upon agriculture. The prospect of new crops; conversions of starchy grains to alcohol for motor fuel; safflower oil, soya bean oil, castor oil as basic chemicals was intriguing.

This led to the development of the first commercial safflower crop and first safflower processing plant in the United States, soon to be followed by an attempted industrial power alcohol project. The Chemical Foundation, through Dr. Pijon, John Foley and Ray Norton, put me in touch with Dr. Leo M. Christiansen, then the Director of the Chemurgy Department of the University of Nebraska. Through Dr. Christiansen, I met Dr. William J. Hale of Dow Chemical Company and brother-in-law to Willard Dow. From these men, I heard of the part that Henry Ford played in the beginning of the chemurgic concept. From all of these men I gleaned my concept of business, agriculture and economics. They passed on to me a valuable understanding to which I have added both success and failure over the past years. The results make up what I am, who I am and why I think as I do. It now becomes my job to pass it on to others who want to know "how to win with what's next!"

There are young, excited research people in various biomass study groups in various universities. These people could get out in the world and build some of these new industries. Those of us of the thirties and the forties were born thirty years too soon. We were privileged to live in the great expansion time of ridiculously cheap fossil fuel. Only now are people ready to change to new energy and new product sources.

New crops have been slowly introduced to the agricultural community. There is a great reluctance on the part of the farmer to change from the old and the familiar. The experience with safflower proved that it can be done, however. Much more work should be done investigating plant life that would adapt to both food and industrial uses. Somewhere a new plant is available in wild state that could be genetically improved to produce and make the term "energy farming" a reality.

Industrial uses of crops seem wrong to many people. They believe that all of agriculture should be used for food. This is not possible. In the first place, only a small portion of a crop such as corn or wheat is directly usable as human food. By far the greatest tonnage is in hulls, shucks and stalks. In the past there has been a continuous oversupply of starch in the world. Therefore, when the best nutrients of the grain are removed, the part that is left must be used for other purposes.

One of the most important uses of cellulose is to convert it to a type of food that is most scarce in the world — protein. The most logical converter of cellulose is the ruminant animal — cattle, sheep and goats. Most people think of grain in connection with these animals. Nothing could be further from the fact. The cow, for example, is an automatic harvester designed as a fermentation machine. As this type of machine, it is a tool to convert grass, straw, hay, milling waste, weeds, weed seed, recycled manure, wood chips and other non-edibles to a most desired form of protein food. Only in the few final months prior to slaughter is it grain-fed, and then only as a part of the ration. One of the things that all cattlemen learn early in their careers is the fact that uncontrolled use of grain will kill their cattle. There are other methods of bringing cattle to market condition besides using grain.

One fact must be realized. That is, just because there are mountains of surplus grain in the big grain-producing countries, it does not follow that the hungry people will have full stomachs. It costs a lot of money to produce a hundred pounds of grain. The farmer must be paid or he goes out of business. Someone must buy the grain, ship it and distribute it to the hungry if they are to use it. Who is this to be? The taxpayers of the United States have spent billions upon billions of dollars for this purpose, as have many other nations to a lesser degree. The mountains of surplus still build up and the hungry are still hungry.

The North American Continent can greatly increase its production of food. Millions of acres of land in South America,

Africa and Asia have yet to be put into cultivation. The cattle industry hasn't begun to utilize the grasses and cellulose material that could produce meat. Many so-called experts on population and food are off base in their views on the limit of the food supply.

The main problems of the starving people is their lack of a way to produce for themselves because of their political and economic systems. There are very few countries that could not feed their people from their own land and waters.

A typical example can be found in India. They have millions of starving people. They also have one of the largest cow herds in the world. Many do not eat meat because of religious beliefs. Anyone who will not eat the meat of the ruminant animal is depriving himself of the only source of "people food" from over two-thirds of the land surface. A small percentage of the world's land can be properly cultivated. Conversion by the ruminant is the only way people can eat grass.

We cannot eat all the available starch and cellulose in the form of meat either. Great amounts can go to chemical conversions and be used in plastics, wall boards, adhesives, pharmaceuticals, coatings, solvents and energy fuels.

A few examples of "things to investigate and do" with crops and residues follow:

Ethyl Alcohol can be made from starchy grains by fermentation. Making alcohol in this manner was one of the first chemical industries known to man. For some reason, he learned to make wine very early in his development. The technique has changed very little for centuries.

Alcohol has a place in motor fuel as an additive to gasoline. For each 1% of alcohol added, the octane rating of the fuel is raised about one point. A 10% blend in pipeline-grade gasoline will replace all the tetraethyl lead and is a quality fuel for modern engines. Being a high torque fuel, it gives a long stroke piston more "lugging" power. It does away with the highly poisonous lead that is exhausted into the atmosphere from leaded gasoline.

In the past few years alcohol-gasoline blends have become

available in many areas of the United States. Many new production facilities are being planned to manufacture alcohol. However, the most important development has been the enterprising farmers who have erected stills on their own farms. They can produce fuel for their trucks, tractors and cars at very low cost. They can also produce fuel to be used without blending with gasoline.

Those who might consider themselves econergists could be most helpful by supporting the farmers in their effort to be free to produce alcohol without regulation. Every gallon each farmer produces reduces the need to import a gallon of foreign product.

Methane is another gas that is obtained by fermentation. This is the simplest form of obtaining an energy gas, and it works best in smaller digesters. Anaerobic digesters in every farm home would be most practical. They operate on waste paper, grass clippings, garbage, trash, sewage, manure, food processors' waste and other organic problem materials. The gas is clean and odorless and can be used wherever natural gas would be used. The residue from the digester is an ideal fertilizer. When recycled and blended in a ration, it makes good cattle feed. Gas production varies with material used and bacterial growth but, as a rule of thumb, one pound of organic material can produce about ten cubic feet of methane. The time required for the batch process per digester is about one week.

There is research being done on anaerobic digesters in almost every university. In a few years, much more will be known about this simple yet little-developed process. It is, however, an ideal project to develop *now*. It is so easy to do that the "Model T" technology is practical. For those who would enter the gate of the econergy frontier, I say, look at methane. Individuals and small companies can compete here. Hundreds of these units are in use on small farms in India. They are practical anywhere.

The manufacture of methane is the simplest to do; it is applicable to the most people in the most geographical areas of the world; it can use the most diverse renewable materials of any present process; it changes present waste products to usable nonpolluting products; with proper instruction and initial supervision almost anyone can operate a digester; the methane and fertilizer/cattle feed residue is usable with present equipment and production practices; it is the most applicable to the individual and small farm or factory use; new or small manufacturing companies can make digester units for sale to users. The university is the best source of information.

Cornstalks are important enough to be specially treated. They are found in most agricultural countries and are especially abundant in the midwestern United States. To point out how important this one energy source is, I will refer to a most interesting study group at the University of Minnesota.

I recently ended an eighteen month project in Minnesota. While there, I was honored to be allowed to meet with a University group of research people known as the Bio-Energy Study Group. The groups meets monthly in the Space Science Center and reports on progress of various research projects on bio-mass. There is a joint effort by almost every science department; including chemical engineering, mechanical engineering, agricultural engineering, soil science, animal science, botany, agronomy, horticulture, civil engineering, bio-chemistry, forestry and space sciences. Outside members attend from various government agencies.

As an econergist, I participate in such groups with great interest. The science and technology are in excellent hands. However, I would like to see the Business School of a university organize and lead such a project with the main research done on how to promote the projects into going businesses.

Many other groups do research on bio-mass at various universities. But this one is the one in which I have firsthand knowledge, so I will report on it.

The first thing that I noticed when I attended the first meeting was the capability and the dedication to the task by this university group. I also noted their pitifully small budget. One would expect an effort as important as this to have a rather large amount of money available. I noted great patience on the part of these researchers who are most accustomed to dealing with government funding agencies.

But the problems of university finance being what they are, department heads long ago found ways to find the ideas, find the people to do the work, find the equipment, and with no visible means of support, to fund and complete research.

The result of their work in 1975 is reported in "Recovery of Energy from Farm Solid Wastes and Timber Production Residues," prepared for "Minnesota Pollution Control Agency, Division of Solid Waste," by "The Center for Studies of the Physical Environment, University of Minnesota." The report covers five projects: (1) Effect Upon Soil Properties of Utilization of Plant Residues for Fuel Energy; (2) The Development and Assessment of Strategies for Firing Crop Wastes in Minnesota; (3) Pyrolysis of Crop and Forestry Residues; (4) Anaerobic Digestion of Crop and Forestry Residues and Manures for Methane Generation; and (5) Energy Recovery through Anaerobic Digestion of Animal Manures.

This is a good report because it does not try to present any of the projects as ready for commercial application. It is a basic study to analyze fact and to prove the feasibility of using the facts in a commercial venture. Universities do this type of basic research best. It is the job of industry to do development, pilot plant testing, production and marketing.

One section of the research has to do with burning crop wastes in the present coal electric plants of Minnesota. It deals with the simple fact that cornstalks will burn the same as coal and perhaps with less pollution. It also points up the fact that present plants can burn renewable fuel.

The nature of the study made it mandatory that the project consider the cost of collecting cornstalks and grain straw. The resulting computer cost analysis is of importance to all industry that plans for the use of cellulose as a fuel or as a chemical base. Most livestock owners and farmers should have this study since it considers many alternative methods of handling in each step of precollect, collect, retrieve, transfer to on-farm storage, storage and processing for use. This feature in the total study will be of great use to those who use any crop in a

23

factory program.

The other study in the report is that on "Pyrolysis of Crop and Forestry Residues." This finding supports the idea that the application of mechanical grinding and the proper application of heat to organic residues will produce two important products, char and tar. Char is a co-product that can be used on site as fuel to produce a part of the heat required in the process. The tar is a complex liquid material with a consistency approximately that of a number six diesel fuel. It is a chemical feed stock for various chemical processes and perhaps as a motor fuel. The process of shredding, heating and separating the char from the tar was proven on a laboratory scale. It is ready as an idea to be picked up by industry for further investigation, a pilot plant run and commercial application.

Could this be the one ready for a small midwestern company looking for a big idea? A way to find out is to try it.

GREENHOUSES

Everybody knows about greenhouses. But have we considered all of their uses? Have we thought of them as new tools to be used in the new economy that will be created as our minds change to ECONERGY — THE CONCEPT OF PLENTY? Are there new industries to use greenhouses that can be developed now? The object here is to point out how many things there are to do in almost every community. Most of these things are simple and can be done by those who would improve their lot — simple things to develop — like greenhouses.

It must not be overlooked that a greenhouse is a solar collector and that the heat collected by day can be stored in rocks or water for use at night. As such a unit it works well as a heat source for a home and it will produce all of the vegetables for a family. But it also is the basis for several local or regional businesses.

New uses for greenhouses in large numbers can be to produce food; to produce crops for chemicals and energy; to do research in new crops, genetics and photosynthesis.

Since food is the first priority of people, many units can be built especially to produce the highly perishable foods. The production of these foods could be the basis for a new vertically integrated industry for almost any community. Since the production from a greenhouse is fairly constant the year round, a freezing, canning and storage business could be made a part of the complex. These additions then suggest the need for marketing and distribution. The community would then enjoy the impact of new jobs in raw material production, processing, marketing and transportation, and perhaps some spin-off industry from waste material.

A by-product of this operation would be cellulose from stalks, hulls, rinds and processing waste. Depending on the circumstances, these could be converted to methane gas and fertilizer, or could be converted to the char and tar described above under "Cornstalks."

Chemical crops and energy crops could be an ideal use of greenhouses. An example of a chemical crop might be the feasibility of growing a crop like castor beans. They are grown as a field crop today but they do not fit in well on the farm. The bean and the leaves contain an alkaloid that is poisonous to animals and people. The implications of pending disaster are evident should children eat the beans or if cattle should break into the field. But the oil is a very good chemical base for many skin care products, solvents, plastics, lubricants, etc. The stalks, leaves, hulls and protein meal have by-product uses.

Dr. Leo M. Christiansen told a story about castor beans many years ago. It seems that they arrived in the United States in the nineteenth century with a colony of German immigrants who came to Nebraska by way of North Africa. They had left Europe and planned to settle permanently in Africa.

While there, they noted the strange religious customs of the native people who would plant a row of castor beans around each of their fields. They were told by their priests that the beans would ward off the evil spirits that brought the plague of locusts to destroy the crops. The German colony soon learned to do the same. They also would plant a castor bean by each outside door. The result was that there were fewer locusts eating the crops and fewer flies in the house. The poison in the leaves tended to make the insects groggy, thereby making them easy prey for birds, and it killed some insects outright.

When the colony moved to Nebraska, they brought the beans and the custom with them. The large families with ten and twelve children established the crop as a commercial venture, and Dr. Christiansen reported that it was the beginning of the castor oil business in this country.

The large families were necessary because the old-type beans would ripen at different times. The bean pod is round with a cord running through the middle. Nature provided a way for the seed to shatter. As the pod dried, the cord got tighter. At the point of breaking, it would explode, throwing beans many feet from the stalk. With this going on over many weeks, it was impossible to use a mechanical harvester. Thus, there was the need for many family members to pick bean pods as they ripened prior to shattering.

The families grew up and moved away from the farm. The farms became more mechanized and castor beans were not grown anymore. When the older folks retired, many moved to cities like Lincoln, Nebraska. They continued the custom of raising a castor bean plant by the screen door to keep the insects away. Lincoln had a lot of these tall ornamental plants.

Many years later, a researcher at the University wanted to breed some new varieties of castor beans that would not shatter, so that the harvest could be gathered with a combine. As he drove down a Lincoln, Nebraska street, he noticed a castor plant in late spring that still had all the bean pods on it. He had never seen this before — usually in the fall they had all popped off.

A little lady who lived in this house had, by accident, developed a non-shattering variety by selection. Year after year, as spring came, she would go to her last year's plant and find perhaps one remaining pod. Over the years, by doing the same thing over and over, more and more pods stayed on the stalk. Today, the beans can all be harvested at one time, mechanically.

There are many types of plant life that would fit the controlled climate of a greenhouse. Some of them may produce more tar than others. Some may grow so fast that char and tar would be feasible as a principal product. Many plants may produce a rubber-like substance that could be a possible replacement for petroleum in tires and similar products. Some may produce organic material not yet considered as raw material for products to replace some of our metals. It is time to channel our minds and our research in this direction.

But, greenhouses in northern regions are consumers of energy, and as such should not be connected to the natural gas systems. There is one location, however, at which they will have all the heat they can use, and also fill a great environmental need. That place is next door to every coal-fired electric power plant.

Hot waste water from these plants is under sttack as a pollutant of streams and lakes. Even more reason to use the water for a productive purpose is economics. When a power plant uses three pounds of coal to make electricity, one pound is converted to electricity and two pounds are lost in the hot water and stack gas. This means that we get only about one-third of the energy value from coal.

Nearly all utility companies are considering better use of the energy stored in the hot water. Heating and cooling of buildings is done when the power plant is located near a market. It is a problem to most of those plants farther away from large buildings. Heating greenhouses would be feasible.

One problem with growing crops by utility companies is the fact that they are industrial-oriented in management and operation. They don't feel confident in the strange world of growing things, and may be less confident in the marketing of food. However, they are in the business of selling energy and as such might welcome a proposal from a new industry group to test and produce near them and to buy the hot water as an energy supply.

Growing things for food, chemicals and energy in greenhouses may find big company help to small companies which desire to start in this field. The prospects are so great that every community with coal-fired plants could have a project.

The spin-off businesses are also great. For instance, after doing some of the things mentioned, how about running the cooler water through a fish hatchery, then to growing ponds where fish production would be year round? They could feed on the by-products from the greenhouses.

So maybe greenhouses aren't so dull or so common.

RUBBISH AND GARBAGE

Rubbish, garbage and processing waste are a growing concern in every community. As the "affluent society" becomes more affluent, rubbish volume increases proportionately. Places to dispose of this material are now hard to find.

Landfill projects are the most common. One would think this to be a simple solution, but land near cities has better uses and the danger of underground water and stream pollution is great.

It has been proposed that recycling the material of the rubbish would be the answer. This, too, has its economic limitations since most of the material is organic and subject to decomposition. A few metals that are in large pieces or in large numbers, like cans, might be economically recycled. It is doubtful that more than a small percentage of the trash can be so handled. This leaves the greatest amount to be used in other ways.

In addition to the problem of where to put it is the problem of custom and, unfortunately, politics. There are many businesses in each city that are built around trash pickup and disposal. Those businesses are usually controlled by permits issued by the city, county and state. This feature of "permits" tends to grant certain exclusive rights to a business that can be quite profitable. A business of this type does not welcome change; therefore, those with the permits and the politicians work hard to maintain the status quo. On the surface, there is much talk and a running-in-place type of action to solve the problem. The problem does not get solved.

There are two things which at this time seem the most obvious and the most ready in the way of new business to solve this problem of waste disposal, and also make more energy and more food available. It seems that the project could best be done by private business but the political facts of life may make municipal ownership a practical approach. It is not necessary to debate this question here. The object is to get someone to do something.

One very practical approach is one made in Jamaica. They have the same trash problem as everyone else. They also have a need to heal the heavy scars caused to the land by the bauxite mines. A private group set up a simple operation that solved both problems. They had hoped to expand the project to the major cities of the United States but some of the political problems stated above slowed such growth.

However, the system is practical. First, the trash is hauled to the processing site. The major large pieces of iron and steel are removed. The remainder is immediately ground. After the first grinding, it is reported that there is no objectional odor.

All of the cans, metals, wood, paper, glass, plastic, etc. is then ground again and placed in large windrows under open-sided, roofed buildings. The windrows create a compost pile in which the material decomposes. It is turned and thus mixed and allowed to complete the composting.

After completion of this process, it is again ground and the resultant product hauled to the open pit bauxite mines as a soil cover to support new plant life. It was found that this material was superior in plant food and contained just about all the nutrients, minerals and trace minerals needed in agriculture. The product looks and feels like peat moss. Therefore, some of it was bagged and sold to a second market — greenhouses.

This is a good way to cycle and recycle rubbish. Its final use is in the production of food and cellulose which we can use over and over again. When done on a nationwide scale, the cost of the plant food should be low enough to allow it to compete with any petroleum-base fertilizer.

The second obvious use of rubbish is one that will produce

both energy and fertilizer. As pointed out in the section on methane, organic waste is the best material for its manufacture and the co-product or residue is a good fertilizer. This type of process works best in small units. Thus, various locations in or near the city would reduce hauling distance and the costs of end products. The gas would be produced near its market and the plant food could be marketed within a small radius from the city.

Some cities are setting up systems to burn their wastes to generate power for various uses. Others are seriously looking into the processes described here. There are needs so great that many new people are required in the field. By pointing it out someone may think of a new idea and get up enough courage to do something about it.

And there are such people in this country. A group in the southwest is drilling into old landfills in a similar manner that oil and gas wells are drilled. They are tapping the methane gase that is trapped in the old fills and selling it to a utility company to add to natural gas.

Is there any doubt, then, that methane gas can be deliberately manufactured in a local digester?

FLYWHEELS

Kent's Mechanical Engineers' Handbook of 1938 states: "The function of a flywheel is to store up and to restore the periodic fluctuations of energy given to or taken from an engine or machine, and thus to keep approximately constant, the velocity of rotation." Those who have rolled an auto tire down the street have seen this principle at work. A few slaps of the hand on the tire keeps the tire rolling and demands that the person trot alongside at a good gait to keep up.

The early one-cylinder internal combustion engines used the flywheel to keep the crankshaft turning while the piston was on exhaust stroke, intake stroke and compression stroke, to then fire and again store the energy of combustion in the spinning wheel. As cylinders were added, the flywheels became smaller but they are still used to deliver an even flow of power.

To understand that other forms of energy can apply to the flywheel takes little imagination. Any force, especially an intermittent one like slapping a rolling tire with the hand will keep the wheel charged with energy. This feature could furnish electricity for all the needs of a home with a unit that would be powered from a solar source.

According to the "World Environment Newsletter" of *Saturday Review* (2-9-74), such a flywheel would be five times more efficient than the internal combustion engine and would be totally pollution free. Large units could store up to 20,000 kilowatt hours of energy. Smaller units that would store up to 30 kilowatt hours would run a car about 200 miles at 60 miles per hour.

There are no basic scientific discoveries that need to be made in this field. The principles are well known scientific fact. There is, however, a need to arrange these facts in a new sequence with new energy sources that will be reflected in a new piece of equipment.

Those who should be interested in this type of development are existing small equipment manufacturers, investors, promoters, bankers, cities looking for new industry, research and development people and those who have a desire to make things better for everyone else and for themselves.

There are several companies in the country and around the world which are developing units of this type. Some have units that are ready for demonstration and marketing. But this should not deter those who would enter the field. There are many uses for this type of equipment. There is a place for hundreds of manufacturers. No one company will have a monopoly. It is like entering farming. One should not refuse to become involved just because someone else is in the business. The market is big, big, big!

So you don't know anything about flywheels? Go out and hire yourself a trained, imaginative engineer, or maybe two of them.

So you don't have any money? Go out and get some.

So you don't know where to go? Well, read on. See if the chapter on "Promotion of New Companies" might give you an idea.

So you don't want to do it yourself? Hire yourself a professional promoter.

BLENDING

These eight sub-chapters on new industry prospects — solar power, wind, water, crops, greenhouses, rubbish, flywheels and blending — are but a few of the ideas and project areas to be considered by the entrepreneur. This book is directed to the established business promoter, to those who would like to be promoters, to those who would like to build one new idea into their life's work, and to the average person who can help solve a world problem by merely understanding and supporting.

Anyone who would do this type of work must always obtain the proper mix of ideas, products, people, land and money. It is in this sense that I use the term, "blending."

We tend to be purists and extremely single-minded in our approach to areas of complex choice. We have been brainwashed into believing that no small business can compete with big business, that petroleum is our only fuel, that steel is the best material for strength, that "X" brand is better than "Y" brand and that ownership of certain luxury cars indicates "doing good." Allowing one's mind to be trapped in this sort of thing is the same as closing one's mind. We are entering an era of multiple types of energy and products. No one type will supply all needs in the future. No one type will always work well alone. Many times a blend with something else will be better.

What would have happened to the dairy industry if they had taken a different course in the late 1920's and the 1930's? The industry felt greatly threatened by a new upstart product called margarine. The action that they took was to try to eliminate the product by legislation. Millions of dollars have been spent each year to keep margarine from use by law. The question is, how much better would the butter industry have done if it had spent its millions of dollars each year on research to find ways to blend butter with margarine? Then, for each ton of margarine sold by the vegetable oil industry, perhaps a ton of butter would have been sold as part of the new blended product.

The battle still rages. Now we see the margarine industry giving the "hard sell" against butter on the basis of cholesterol content. This advertising plan may turn around and bite the margarine people. They may be saying the same sort of thing as a wine manufacturer who states that, "wine is better for alcoholics than beer." It may soon be understood that the chemical function in certain people may be such that building cholesterol in the body may resemble the problem certain people have as diabetics with sugar and alcoholics with alcohol. The point is that we should not let our minds become narrow by the millions of dollars of brainwashing-type of advertising directed toward us daily. It does not seem to be an honest business approach for either the margarine people to advertise in this way or for the butter people to legislate against the use of margarine.

There are many examples of narrow thinking. In the cattle business, it took years before crossbred cattle were accepted by the feeding industry. Angus-Hereford crosses were severely discounted in price. Today they bring a premium price because of their heterosis or hybrid vigor. However, there are still those who believe that Angus should remain Angus, and Herefords should remain Herefords, when in reality those who do crossbreed are an extension of the market for purebred producers of each breed. It is just an example of the fact that there is room in all business for some to do it differently. In many cases, the blending of the "different" makes the total business better.

No one person can know everything needed to build a complete business. There is almost aways the features of land, raw material production, processing and marketing to be considered. It may be possible to build a business in just one of these areas. But even so, there is the diversity of minds and backgrounds needed which, in many cases, means clashing rather than blending.

There is no sure way to pick people who will always work well together. If there is a formula, after assurance that the skill and technology requirements are satisfied, the next best thing to look for is a good responsive human being. If a mistake is made in the selection of people, the misfit usually separates himself or herself from the people mixture.

There are many areas of blending to consider other than how to combine product, people and energy sources; the proper blend of yourself with desire to build and desire for money; the blend with what you might want with what the consumer might want; the blend of your skills with the continuing operations of the company.

As inventor or promoter or investor, it is well to know when to let go of ego orientation in favor of the answer to the basic question, "Does this make good business sense?"

When the answer to this question is seriously arrived at, it seldom results in compromise or conflict.

CHAPTER IV

PROMOTION OF NEW COMPANIES

New industry comes into being by changing an idea into reality through hard work, dedication and integrity. These three ingredients are possessed by most people. When properly directed toward a good idea, they are the basic building blocks of real wealth. No other form of potential wealth is so latent or so evenly spread among the people of the earth.

The entrepreneural endeavor known as promotion means different things to different people as it applies to different situations. Herein the term is used to mean the organization and financing of new business ventures. This feature of business is little understood by the average person or within the business community itself. However, there is never a new product, new process or new company that ever exists without a promoter in its development. This is true whether done in a democracy or in a communist country, in a big company or in a small company. Some person or group must present the merits of each project so that it, at some point, gains top priority for funds and talent.

Government backed the space program that took the United States to the moon. The first promoter to suggest this idea was surely assumed to be crazy, stupid, irresponsible, self-serving or perhaps more kindly treated as naive, idealistic or poorly advised.

All new ideas are accepted slowly. Through persistent effort, fact is built upon fact, the idea gains friends, then participants. Soon it is changed from that nebulous thing called an idea to a positive, written plan. At this point, it is possible to add money which, in turn, adds people and hardware. Research, development and testing result in the reality. In the case of the moon landing, it allowed worldwide participation through the medium of television. This was promotion by big government. But the entire trick involves changing an idea into a reality that people can see, feel and use.

Government finances its top priority projects by using the wealth of the tax base of the country. Funding the private company must take a different approach. It must seek out and convince each investor to risk his money. To find those with investment funds and investment risk-taking nerve is a specialized business function.

Real risk-takers are few. A program which presents sound management, low risk and high projected gain may not achieve investment interest. Therefore, in recent years, an additional incentive has been added in the form of the tax shelter.

The expense of an operation, especially when the investment is highly leveraged, has attracted money to tax sheltered business structures in speculative oil exploration, real estate, cattle breeding and feeding, research and other high risk ventures. It is planned so that the investment reduces current tax liability and defers the taxes to a future year when the investor is in a lower income bracket, or when the investment may be sold as a long-term capital gain with a lower tax rate.

Often in the past, promoters have mistakenly built new ventures for the sole purpose of the tax shelter. These businesses seldom succeeded because there is no such thing as a "tax shelter business." There are only businesses that, because of their expense structure, can defer income to a future date. To mean anything to an investor, they must be viable, profitable entities that will return original cash plus a profit. Therefore, the first consideration is a meaningful, well-managed business that *may*, as a fringe benefit, defer taxes. As such, tax benefits should be structured into new ventures when applicable.

The intricacy of the tax work that is implied here requires that experts in taxes become involved with promotion. But there are other specialties required if money is to be raised outside the promoter group. There are state laws and federal laws that regulate the process of selling security instruments (stocks, bonds, partnership interests, participations, etc.). Some of the new regulations expand the definitions of securities, and it is not wise to assume any untrained reasoning on this matter. Therefore, an attorney with a security specialty must be made a part of the new effort. The list of specialists becomes longer as the complexity of the business increases. It is quite evident that promotion of new ventures is most often a team effort rather than a one-man endeavor.

Many errors may be made by the inexperienced who have good new ideas and wish to convert them to a going business. Not always is the good idea in possession of a person or a small company with the resources and the know-how required for development.

A common mistake is to assume that a bank will loan money on the idea. This is seldom possible unless the person or company applying for the loan is strong enough financially to support the project by other income or assets. About the only call that should be made at a bank would be one to find a reputable promoter. Once in a while, a banker may know of someone who will help in such matters.

Some certified public accounting firms and some business attorneys know the people who do the specialized work. They are not easy to find. The most capable ones are often busy on their own projects. Most of them welcome a look at a new project if it is in an area of their particular interest. They seldom advertise, although I would suggest that they should. Many people need their services, especially new entrepreneurs with new ideas in the field of econergy.

It is usually an error to expect security dealers to take an idea and convert it into a company for which they will raise the capital. There are some firms in the investment banking field who do this. However, they usually will look only at established public companies with an earning record. There are exceptions to everything but a new untried business will have difficulty finding an established security dealer that can help.

Funding new enterprise with debt is a method that is not usually possible. The reasons for this are obvious when seeking the loan from institutions but there are a number of methods which make this possible from other places.

As an example, it may be necessary to assure a new investor that he will get his money back before the profits are divided among all the other owners. In this instance, the investor puts up the money with a form of investment that pays him interest, gives him a secured position on the assets and returns his original investment from the first profits. This system works best in ventures that have not had large pre-organizational expense or investment. It is a simple case of the know-how taking an interest in the venture and working for a salary until the money is returned. Then both parties participate in the rewards of the enterprise from that time on. This is one of the few examples where debt is really justified on a long-term basis for a new and untried business — debt with payment tied to profits when earned.

The surest road to success is for new developments to look for equity as the basis of capitalization. This simply means going outside one's own resources and adding funds from outside people. In other words, allow the same people that fund the money lender to participate fully and directly as owners in the new venture.

To this point you are no doubt wondering if debt is to be used at all. One being leery of debt must also admit that debt, properly leveraged, is one of the few ways to multiply investment rapidly.

Most new companies are cash poor and the debt that they have will force them out of business on the first major business reverse. They are not only leveraging their business assets but also the fact that they hope to always make enough profit to service debt.

It is much easier and much safer to leverage against a cash

position because both your company and your lender can deal on the basis of strength. The company has staying power and the officer of the lender can stand in front of his loan committee and point to the cash position with confidence.

With this basis, loans for new business are justified. In fact, with proper cash positions, 100% borrowing is practical. It is also possible with proper cash position to never spend the money as your company grows through merger and acquisition.

Therefore, raise enough money for cash position to attract and protect all you need to borrow for plant, equipment, inventory, receivables, etc.

Many communities will build and lease back to industry in order to create new jobs and a new tax base. They usually do this by selling tax-free municipal revenue bonds. Where a large building is required, this may be an ideal way to gain the use of capital.

Some areas have existing buildings available that may be of use with some remodeling. Many cities, both large and small, have purchased old military camps, old military ammunition depots or old airports. Small new companies can get a start here, and usually they find that they will have plenty of room for growth.

Often the areas most suited to the new econergistic industry are the small communities. Some, of course, should be developed in the large metropolitan centers but most of the ideas in this book have to do with the smaller, less complicated ways to accomplish this concept of plenty. Many of the projects can be started for a capital investment about the size in a family farm. This is not an amount that a rural community would find hard to fund. It is doubtful that there is a county in the United States that does not have thirty or thirty-five citizens who could and would pool and risk this amount of money for a worthy industrial development. The problem is one of proper structuring and adequate communication with the community.

Many technical people with high academic degrees and in-depth experience now live in small rural towns. Many more would move there if there were a job in their field, just to have a peaceful life for themselves and their families. The labor force is usually adequate for the average small plant.

These features may indicate that a turnaround could be in the making. The new industrial movement may not be toward the multinational giant. Much smaller companies could be more efficient in converting the renewable-type raw materials to energy and products. Economy of scale may be offset by less freight, cheaper local energy, local worker pride in their company's product quality and regional acceptance of product by consumers.

As we discuss new companies and their promotion needs, it is wise to consider another approach. Always ask the question, "Does this project or product warrant a new company or would it be a better project for an existing company?"

Often it is best to promote the project as an expansion for a company that already has an operating base. The advantage of having a management team in place is great. It is much easier to add to an existing team than to build a complete and untried new team.

There are many small struggling companies that need only to be reshaped or redirected to be a success. A new econergy project could make them great, especially if it solved their most probable problem — that of being underfinanced.

One who would promote a company or an expansion anywhere should try to keep in mind the basic divisions that make up the parts of the chosen business. Most are made up of four: (1) land, (2) raw materials, (3) processing, and (4) marketing. A business does not have to be involved in all four of these but great care and thought should be given to being fully "self-contained" if there is any chance of undue pressure from outside sources that would break control of raw material, product or market. It is always better to spend more time for adequate funding of a more complex company than to build a small one with less capital and with little control of its future.

When the company is at the point where outside money should be brought in, it is then necessary to build the financial package with features most desired by investors. These, in general, are: (1) safety, (2) growth, (3) profits, and (4) liquidity.

When these four features of an investment are fully explored and properly balanced the company has a pretty good chance of presenting a satisfactory investment package.

Safety is of prime importance to investors. It implies many things, but in general it is concerned with those things that will assure continuous operation and value. The one thing which usually reduces safety is debt. Mortgages are contingent conveyances of property, both real and chattel. This feature of debt puts continuous jeopardy upon investors and can wipe out their equities in times of economic imbalance. Structuring debt is an important job of new company promoters and managing debt is a major role of management. New ventures may wish to consider full equity funding of long-term assets and use debt most prudently for short-term inventory and receivables.

All investors know that there is no such thing as a "sure thing," and I don't think that they expect a cinch. What is expected is that the feature of safety be fully analyzed in each venture and, as much as possible, be built in. This makes high risk investments possible in such businesses as oil exploration, research and development and econergistic projects.

Growth is a requirement in a world of shrinking money values. It is the one thing that draws money from banks and savings institutions. There are many safe places to put money to return a reasonable annual interest rate and hold the money together. But take the case of the person who put $10,000 in a savings account ten years ago. Just about any investment in real estate, products or commodities would have done better. The 1970's has seen a great crisis in the value of money. The 1980's may see this crisis continue.

There is, however, another type of growth which is not one of keeping up with inflation. It is one that grows because of profits and expansion. This is the type generated because the business entity is producing more than it consumes. It may be the most important feature of an investment. It is important that a new business has the feature of growth if it is to attract new investment money.

Profits are the main reason to promote a new business.

41

Without profits, a business can exist but for a short time. The idea of profits has been the subject of much criticism. Profits give some people a feeling of guilt because they believe they were the result of charging too much for a product or service. Some refer to them as the "degree of greed." Some believe that the communist system is better because profits are not a part of their business purpose. This is not exactly the case.

A profit indicates that an enterprise produces more than it consumes. As such, it is the only way the communist or the capitalist enterprise can survive. In the typical communist country, when item "X" sells for one ruble and is produced for one-half ruble, the state takes the one-half ruble profit. In the capitalist country, when item "X" sells for one dollar and costs fifty cents to make, the state takes a big part of the fifty cent profit as taxes and leaves some for the investor. So, profit is produced in each political system.

Profit often comes after the feature of growth in importance because of the tax structures. Growth is not taxed until it is realized as a long-term profit. As such, it is not usually taxed until the investor sells the investment, thus giving some control as to when the tax is due.

Profit projections should indicate strong profit margins for newly promoted industry. The ones with small margins may not interest new investment money. If they cannot do better than the old businesses, why should anyone change their investment portfolio?

Liquidity of some type is necessary where more than one investor is involved in a business. Individual circumstances change daily. It is quite possible that some of these changes will be such that a sale of an investment is necessary. It is very important that liquidity be available to all investors in some form.

This does not necessarily mean that the investment be liquid on an immediate basis. The listed stock markets try to do this, but they often give up real value in favor of the immediate sale. New companies will not have listed stock exchanges available to them in their early years. Smaller companies that we are considering here will do well to solve this marketing requirement on a little longer time element and within their own resources and marketing capabilities. The idea is to give it meaning and deliver results.

From the discussion so far in this chapter, one gets a picture in the form of a jigsaw puzzle as to what a promotion looks like. And that is exactly what it is. It is an idea made up of a lot of parts that have not yet been put together. As indicated, there are people who do this "put together" as a business. There are good ones and bad ones, creative ones and uncreative ones. They are usually busy people because there are so many new things that need to be done. This is why these people are hard to find. They are buried as officers in the activities of some company and are seldom referred to as promoters. Therefore, those with the new ideas should consider being their own promoter.

As previously stated, being a promoter is usually a team effort. The process of building a stockpile of facts and parts should also attract other people. Among those other people should be one who reduces the jumble of facts and parts to an organized, written document. Talking about a new idea does not convey the real project to many people. Others can only relate to their own mental pictures that are limited to their own experiences.

It is like an architect trying to describe his idea for a new building. It doesn't mean much until he has drawn a picture on paper. Then, as detail plan is added to detail plan, it becomes a total plan that any knowledgeable contractor can execute. Often the one with the idea for the new building is not the architect but one who can draw a rough picture from which an architect can develop the finished plan. All the one with the idea needs to do is to contact the architect.

But how can one change an idea into a going business? There may be few definite approaches that would fit all the new promotions of new enterprises that must be developed in the immediate future. That would be like setting up a definite color scheme and drawing for each picture that an artist is to paint. New industry, like new paintings, is creative in nature so that each is different from the other. However, there are certain phases that all creations of this type go through. Those can be described, and the special project circumstances can be categorized into these phases.

1. *The Creation Phase* is the most remote. It is locked in the lonely hours and the quiet of late night. It is the phase, in its early stages, that must be the stroke of genius reserved for only one. Then the sharing of ideas, the brainstorming of the creation to the point of practicality can involve many minds.

2. *Documentation Phase* is a mechanical and work phase which converts the idea to the written plan. It is here that those of the creation phase can read back their first thoughts and analyze them for completeness and practicality. Here, too, is where the ideas and concepts begin to pass to others whose minds should be concerned with cold, hard facts and structuring.

3. *The Research and Development Phase* is one in which known facts are arranged and rearranged into new sequences, new processes or new products. It is not designed to be the phase of basic or pure research. It is the commercial pilot plan stage to which research and development is also applied to the new requirements of corporate structuring, finance and marketing.

4. *The Production-Marketing Phase* is the fruition of the idea. This is the final step in which the planning proves profitable or unprofitable. It is the goal of the effort of converting the idea to the reality. It is a balance of making and selling. It is the time of greater affluence. It is the time of production of wealth.

Many people see new things and have creative ideas on new uses of old things. These ideas are what the industrial world needs. The following is an example of creation phase thinking.

It might be practical at this point to discuss an idea that is still in the early creation phase. Consider a crop that I saw harvested from the sea in the Cayman Islands, British West Indies. An investor group had built an experimental-commercial business designed to preserve the green sea turtle. The products were turtle meat and the co-products of the turtle shells. One of the new things to be learned was how to harvest turtle grass to feed to the confined turtles. It was this grass that really was eye catching as a possible new industry for the Caribbean and Gulf of Mexico shore lines.

Turtle grass grows in the shallow water of the estuaries and coasts of most of the Caribbean. The turtle farm at Grand Cayman developed a mowing machine on a barge in which the depth of the mower can be adjusted to the depth of the water. As the barge floats over the turtle grass the mower cuts the grass under water. It floats to the surface, is windrowed by the wave action, and loaded on the barge by a screened overshot boom, to then be hauled to the turtles and fed.

It is reported that the grass is about fifteen percent protein and is rich in the nutrients and minerals found in the sea. About ten crops can be harvested each year from the same acreage and the mowing appears to make the growth thicker.

This "sea farming" operation indicates that the crop could be dried and pelleted and shipped to livestock producing areas as a protein-mineral supplement. It would be extremely valuable as a supplement on the grass pastures found in the tropics and subtropics. These pastures are fast-growing and lush but lack protein and mineral nutrients.

Such a feed processing plant could expand into the field of human protein-vitamin-mineral supplement and perhaps some development in pharmaceuticals. The project lends itself to an immediate market in animal agriculture and has this future value which a profitable feed supplement business could research and develop.

This is as far as the investigation has gone for turtle grass as a candidate for a new business. It is not known where an in-depth investigation of this product of "sea-farming" would lead but it does present possibilities. Let's not assume this is ready to go but we can use the idea as an example.

The above story is pretty early in the creation phase. Some better facts to confirm earlier impressions should be obtained but following this investigation several experts in animal nutrition should be brought into the project to start the documentation phase.

The documentation phase will investigate in detail all the factors and facts that can be obtained about the crop. It may include all the technical information available from governments, universities and private industry. This may include information on how to grow, how to harvest, how to process and how to market. All present known uses should be studied and all previous successes and failures, if any, should be documented. If all of this work can be formulated into an operating plan for a new industry and if the pro forma profit and loss looks promising, plans for the research and development phase should be made for the purpose of justifying an expenditure for commercial operation.

In the research and development phase, pilot plant work should prove the operation on a small, low-risk scale. This phase will not only prove the operating procedure but should be the time for developing personnel; for designing the corporate structure of the new company; for developing the investment capital required to build the plant and for operating money; and for working out the marketing and sales program for all the initial production.

Should all of these early pre-operative jobs be completed and the feasibility is still "go," funding and construction of the new venture can be started.

(I didn't say it was going to be easy; I just said that it was important to do things like this. Those who would be promoters of new ventures have much to do. There are few easy shortcuts.)

There are, however, some moral considerations and some understanding of what real truthfulness is, if one is to consider being successful at this business. It is not only that you will be placing at risk other people's money, but also the time and reputation of highly trained technicians and businessmen who can be ruined with scandal and misdeeds. For the purpose of this discussion, a short statement in this regard will suffice.

A great lie may be found in a half-truth. So tell all of the truth, both good and bad, as you present the case for your project. But also know that the worst lie of all is the one that is never told. It is found in the fact of total silence. It is allowing wrong impressions to continue by your silence. It is allowing wrong assumptions or no assumption at all to exist when the bare truth is so simple and strong. It is a job for basically honest people and none others should be involved.

Promotion can stand success and it can stand failure, but it cannot survive dishonesty.

CHAPTER V

ECONERGY AND FOOD

It has been said that the best hiding place is one right in front of the seeker because he will be most unlikely to look there.

Food is the number one priority of mankind. When he has none, he spends all of his time looking for it. When he has plenty, he can't understand why all do not have plenty. He writes and tal talks about it a lot. The more he writes and talks the more evident is the fact that he doesn't fully understand its source, its uses and the distribution-marketing complexes required to feed everyone.

Food is plentiful. Food is everywhere. So why are so many people starving? Why do those societies that are busiest producing other things have the most food? Why do countries like the United States have the capacity to produce more than they can consume with less than five percent of the population involved in production? Why is an industrial country like Japan well fed when it must import raw material, energy and food? Why do farmers and ranchers around the world reduce the size of their beef herds for lack of market when so many need protein food? Why is there a real danger of price-depressing grain surpluses? How can a world with so much furnish some people with so little?

The questions indicate answers that will run counter to most books and articles on the subject of food. It would require a book, at least, for each question to elicit a proper answer. But those who give it thought and write an answer would at some point say that the underfed and under-privileged should be educated. With this we would all agree. However, the education of these hungry people is not possible until we have a broader understanding ourselves about how to use the organics from which our food comes. It will not be possible to transmit an understanding of a socio-economic system that will produce abundant food, clothing, shelter and the jobs and businesses to make food purchase possible until we in the developed world have learned to use the organics and other renewables to supply our own needs in new jobs and the production of wealth.

The purpose here is to fit the econergy concept to the production of food and especially for production in hungry nations. It is hoped that by understanding that food is nothing more than stored solar energy, it is also the source of clothing, shelter and energy. The more we produce for these secondary uses, the more food is available. As more industrial use is made of organics, it creates more jobs, more capability to buy food. Therefore, the farm should be considered a raw material producer for industry and energy as well as for food. Once the developed world understands and produces from these sources, it can pass the know-how on to others. This can change world trade patterns, transportation, politics and cultures everywhere — not immediately, but gradually over several generations.

Again, this is the concept of solving one man's problem at a time. The world food problem, like the world energy problem, is too big to tackle in total. As you and I solve our own small problem of renewable raw material and energy, we will be building systems that could easily apply to that one man or one entity in some small needy country.

One must not believe that these countries are devoid of expertise. Each area has its business centers that can be local springboards for new ideas. It is those which are in the best position to spread the econergy concept and processes that will best fit local custom and culture. This is the process of doing the very small things to bring the right kind of change for us and the underdeveloped. It is done by passing the idea from one person to another, from one business entity to another, from country to country, and from culture to culture. This will change the starving to the well-fed, the poor to a higher affluence, and all from a state of dependency to a state of self-contained sufficiency, especially in food, clothing, shelter and energy.

People in this world are not hungry for lack of food availability. There is enough food for everyone and much more can be produced. When there was so much surplus wheat in the 1930's that a farmer could not trade a bushel for a postage stamp, the hungry of the world were as hungry as they are now. In the mid-70's the need for high protein food was greater than ever and the livestock industry was reducing its cattle herds because of the lack of market. What is evident is the fact that *need* is not always reflected in *demand*. The reason for the lack of demand against available supply is that of cost.

The hungry cannot generate an economy that will furnish them with businesses and jobs so that they have money to buy food. To generate such an economy requires education, change in economic systems and change in political thinking. Since these things are hard to do and take decades to accomplish, food relief programs are the short-term result. These tend to increase the problem since it is among the hungry people that population growth is highest and the perpetuation of ignorance is greatest.

The answer to the problem seems obvious — just have these people produce their own food. But it is not that simple. All the hungry cannot be employed in the jobs of agriculture production and related service industries. The agricultural sector cannot expand, either, without the jobs for those who will make the market. So we have the old problem of which comes first, "the chicken or the egg?" The way to break this dilemma is to back away from the pressing problem of food and look at the organics as raw material to convert to industrial products.

One reason that jobs do not exist is usually a lack of energy and raw material. But most of the hungry world is in regions of abundant plant growth. Learning to use this type material does not require a technology like that needed in fossil fuel production, refining, steel mills, aluminum production, atomic energy or electronics.

As noted in the chapter on New Industry Prospects, most of the things to do are relatively simple and are organic. If some of this type industry were developed, a few new jobs would be generated. With this new income the few new workers can buy food. It is available now to the full extent of their buying power and can be continually available to everyone on the same basis if we can shut off unlimited growth in population. As the new workers grow in numbers, new food production will be made to satisfy their need and demand.

When the present industrial nations were going through the industrial revolution, their education and technology expanded rapidly enough to make most of their people employable. The developing nations were here then but missed out on the chance to participate in the changes the first time around. Their uneducated and unskilled people have now grouped together around the large cities hoping for jobs that will lift them from despair to a few promises of a modern world. They have lost their understanding of how to live off the land. They have also lost their claim to land on which to try.

Those of us who live in industrial countries should not overlook much of the same thing that is happening in major cities with all of the problems for its people that exist in the developing nations — except starvation. Their problems of employment are serious. Their need to work is great. To make jobs available there must be new, small industries that can be built using the econergy concept.

Those people who live in the hunger belt of the world have a good chance to use organics for most of their total needs. Most are in areas of the tropics and subtropics. Most see more stored solar energy in the form of plant life than anywhere else. All of them need energy and fertilizer. All of them can produce it from the plants that they see every day. Their plight is best described in the following story.

Dr. Norman Borlaug spoke at an informal lecture at Colorado State University a few years ago. One would expect to hear

51

about some new developments in plant genetics from his "green revolution" research. Instead, he talked on the need to understand the use of pesticides, herbicides and fertilizer; and on the anthropology of America's Southwest Indian. It was this story about the Indian which makes an analogy with the plight of the hungry today.

The Indians of what is now the Southwest United States were first nomadic hunters and food gatherers. In this arid plains country they were forced to move continuously in search of food. They found it hard to survive. Their diet was very low in protein even though within sight at all times was the world's largest supply of protein food — the buffalo. They did hunt the animal for food and used the hide and bones for shelter, clothes and tools. But since they were on foot, they were assigned the role of being scavengers to the old and sick animals and to an occasional lucky hunt.

One day they saw the thing that was to change their lives. They would move from a miserable, undernourished people who spent 100% of their time gathering food to a robust people with a highly developed culture. The horse ridden onto their land for the first time by the Spaniard was to be their salvation.

By fair means and foul they obtained the horse. They became horsemen to rival the famed Russian Cossack. They were now very mobile. They could ride into the great buffalo herds and in a short time obtain their food, clothing and shelter. What before required 100% of their time now required about half their time. With this free time they developed a culture with organization, art and spiritual philosophy. They were strong, healthy and free. They were self-contained and dependent on no one. Their high protein food supply was almost inexhaustible for their needs — it had always been there within sight but now they could reach it whenever they wanted.

Such is the almost identical plight of the starving millions in the tropical and subtropical zones of earth. Right in front of them is the energy, food and industrial raw material for all they could desire. The "horse" for them to ride is one of understanding — especially to understand the econergy concept.

But to infer that we should just tell them about econergy will not transfer to them any sort of understanding. Those of us in the developed world do not understand it either. We are so engrossed with bigness in our business world that we overlook this new embryo industry. It does not fit the major corporate structure of today. But we need it as much as the poor and starving do. It is up to us in the developed nations to make small starts with our capital, our know-how and our belief in the private enterprise system. It is then our job to "show and tell." Nothing is better than a good example.

To enter one of the countries as an American or European for the purpose of making a good buy on land and to build a farm and ranch business or a manufacturing plant using their cheap labor will not be practical. This would be a form of the old colonial system which is more feared by the hungry nation than starvation. This old system is partly responsible for their ignorance and their hunger. It is time to enter the country as a businessman who will "show and tell" for a profit, but that one day will be gone.

This businessman is a new type of soldier whose war is against ignorance and poverty; who, by winning, will make shooting wars less likely. This contact is most effective on a private basis between the businessman and the entity to be the new production unit. The less government involvement the better, although there is a place for proper government participation. Enough of these one-to-one type contacts between two small entities will start the pyramid effect. The student will at some point become a teacher and one of the most basic laws may well become operable — the law of giving and receiving.

The underdeveloped nation has everything to learn if it is to catch up in food production and in fuilding industrial capacity that fits its culture and geography. One could not expect them to start with modern Iowa-type corn farms. They must first start with a Model "T" type production. The best base to start with is with whatever livestock operations they now have. To add a farming operation to a livestock operation would be a practical start. At least here is a person already in agriculture.

Contrary to most writers on the subject of food, the ruminate animal (cattle, sheep and goat) will be the major source of the high protein diet in both the rich and poor nations. Only a small percent of the land can be in cultivated crops. By far the largest land surface is in grass and plants which these meat animals convert to protein food. They can also clean up the straw, stalks and wastes of the cultivated land. Their meat, when added to dishes made of the grains and vegetables of the farm, make tasty eating and is the basis of larger, stronger bodies and more alert minds.

The industries that these animals create are great. Transportation, processing, packing, freezing and marketing generate the kind of jobs that pay enough to allow the worker to buy more food and more of everything else. Need then converts to demand.

Other plant growth of these regions are possible sources of food, chemicals or energy. Small units are the answer for the start. If they grow big — fine. To start big is folly. This applies to us in the United States as well as to those in the small hungry nation. However, large or small, the manufacture of products, whether as food or for industry, generates wealth with which those who work can buy food.

CHAPTER VI

THE POVERTY BELT — A PLACE OF GREAT WEALTH

There are none so poor as those who have wealth but don't use it. There is nothing more pitiful than those who have wealth and don't know it.

Between the Tropic of Cancer to the north of the Equator and the Tropic of Capricorn south of the Equator is an area of approximately 78 million square miles of earth surface. About one-fourth of this is land; the rest, sea. It is here that most of the poor and hungry live. It is here that most human misery exists. It is here that the population growth is greatest. It is here that ignorance and disease grow even faster than the population.

It is also here that the factors of abundance are present in such amounts that a region of prosperity far greater than that of the temperate zones could exist. All of the ingredients for a life of wealth and comfort exist here. The basic building blocks are sun, water, air and organics. All are renewable. All fulfill the requirements of the econergy concept.

The people of this warm and livable area need only to use these renewable items and arrange them into sequences of production to achieve their greatest dreams of wealth. Perhaps not the same judgment of wealth of the western world, but their own judgment of what makes the good life for them — the easy, slow life, the time for family and fun, the siesta, and a way to produce for a productive economy at a slower pace.

Education and expansion of present cultures can grow as their economies grow. The tropics is a region of great wealth. It is the place to instill the principles of econergy. It is the place that could achieve independence first from the finite supply of metals and fossil fuel. It may be the place that could be exporting products and new technology to the western world at some future date. It is a place of great hope.

Many who have seen the misery, hunger and lack of education in these areas must doubt the speed with which this great hope will be realized. There are so many people who need help — so many beyond the point of being helped — so many more people each year. But it is here that all the parts to achieve affluence are present.

The tropics have the most countries of any area of earth. As a result there is the most diversity in types of government, political thought, culture, religion, economic systems and human need. It is this diversity which may be its salvation. It is, also, the underlying universal need for food and jobs that could direct thoughts and energies toward the concept of econergy. The solutions will not be easy or quick.

There are many conflicting interests within each country and in the world at large which will not support this change in thought toward independence and self-sufficiency in food, clothing, shelter and energy. With political thought and big business thought being what they are, aid from the world industrial powers tends toward building markets rather than helping solve the small country's problem. When help does come it is usually "too big" and "too western."

It may be that most experts sent to the tropics by their western governments or by the world's foundations really do not understand what they see. Most reports on why there is little agriculture in the tropical forest region state that the soil is poor. Yet, as one stands and looks at the production one sees the greatest variety of plants in the world, growing twelve months a year in such abundance that it is difficult to keep a trail open if it is not used often and heavily.

Perhaps this land does not produce the new hybrid grains as well as the midwestern United States but there are many other food crops that do well here. The trick will be to relax our rigid thinking as to what constitutes an economic food crop for today and to learn to use and improve what is already present in the tropics.

The forest area of the tropics is the most efficient converter of solar energy to cellulose, lignin, chemicals and food of any area in the world. For those who will use it properly it will return the needs of the people who live there and in time be a major export area for the needs of those who live in areas of less sunshine and less rainfall.

Look, then, at this most productive region and blend our thinking to the way the tropics really are, and how we can fit it rather than how it can fit us. Not only should we try to understand the nature of the tropics but, also, the nature of the people of the tropics. The slower pace of these people may be what the whole world needs. We in the north may not be ready for the siesta from noon to four o'clock but this much time to relax and think may bring forth more cool and clearheaded decisions for people, business and governments than anything else.

So much has been written about the underdeveloped nations and the plight of their people that many cannot believe that there is an organized society involved in such misery. All of the underdeveloped countries have well-defined business centers, governments and long established trading patterns.

All of us recognize the long established products of the tropics because we use them every day and have done so for many centuries: bananas, coffee, cotton, cacao, cinnamon, cloves, jute, nutmeg, pepper, peanuts, rum, rubber, rice, sugar, tea, tobacco, teak, ebony and mahogany are a few of the common things from the tropics. These and other related products are agricultural and result from the year-round warm sunshine. All of this production has harvesting and processing wastes that can generate food for cattle, sheep and goats, and all will produce energy from fermentation in the form of alcohol or methane.

It is evident that the tapping of this wealth has been and still is one of the great problems in the poorer tropical nations. First came the Colonial system designed to take back as much as possible to the homeland and leave as little as possible in the colony. With that came the plantation which still exists for the same purpose. Then came the emerging governments whose major endeavor was to develop and maintain power. All of these systems tended to drain resources from the producing region and from the people who lived there. As poverty prevailed from one era to another, little was or is now being done to educate and train the new millions of children so that they might be able to turn things around as they matured.

The total problem of ignorance, poverty and hunger in these lands of potential plenty is too big now to solve by one billiant move. The only help must come now from the same basis that the solution to the world energy problem must come. That is to help just you and me to become self-sufficient in energy because our problem is small. The same could apply to a family in the poverty belt. Look to building just one job and it can be done. To try to create a million jobs is almost impossible at this point.

Such an attack on the problem will require government-to-government help, business-to-business help and person-to-person help. In any event it will be necessary to springboard the activity through the local people. Those who plan to do business here should plan to reduce their presence in the future except as an investor, advisor or lender in favor of being a supplier or a customer.

The day of economic control of others is past. The days of doing business together and exchanging what we have in surplus for what they have in surplus are ahead. This will give both parties a chance at abundance and their abundance of sunshine may give them a great advantage in the balance of trade at some future date when they have mastered the concept of econergy.

The methods used now to elevate the economies of the tropical nations must be aimed at generating food and jobs for those who now have neither. Not to solve this problem is to invite the next major war.

The billions of starving people will not stay passive forever. Their great numbers will follow the next great leader who will promise them something better.

Great hordes of people determined to trade their "nothing" for what you have may well be able to do so. Their lack of understanding and formal knowledge will make them great followers of whomever gives them the first hope of their lives.

The next world leader of the billions of people who have nothing need only ask, "Are you getting your share?" The response will be billions of negative answers. This question, asked often enough, could lead to a frenzied march by these billions on the world centers of wealth. The atomic weapons of the world would be useless against such an assault.

As the production centers of wealth were seized by these hordes the wealth would vanish. Such wealth is as fragile as a soap bubble floating on air. Try to grab it and it disappears.

The thing called wealth is supported by a delicate balance of complex factors. The societal factors which allows wealth to exist are far more important to the production of wealth than are the tools of production.

The fact that these people now live in a land of plenty is no more evident to them than to the present unseeing, developed world. It is time for everyone to understand the concept of econergy.

CHAPTER VII

MORE OF EVERYTHING FOR MORE PEOPLE

Nothing can cheapen the value of human life more than the feature of too much human life. Nothing has much chance of reducing present population trends except an effective appeal to the unique human intellect. No human endeavor is more important than making this appeal work. Without its success all other social and economic goals are sure to fail.

We have looked at a new way to think about producing the things that people eat, wear, live in and use in their various cultures. We saw a glimmer of hope that we all have for a life of abundance to be found in the concept of econergy. The problem was reduced to projects that individuals and small groups can carry on for themselves and to those things which the large capital formations and governments can do to help.

It is obvious that we can improve the economics of everyone by conserving the capital assets of fossil fuels, metals and other mined materials that are never replaced once they are used. This point of view does not mean that we should stop using these resources. Such an idea is impossible since the entire world economy is built upon them. The fact is that use will increase and greater effort must be put into finding new reserves of these precious materials.

The job at hand is to start, however small, building systems that will gradually replace these fixed resources so that in the next half century we will have little dependence on them. The problem is a great challenge in light of the fact that in approximately thirty-five years the population of the world will be reaching eight billion people, or about double the present number. There is little doubt now that this will happen, barring great natural disaster, disease or nuclear war. Population experts hope that world population can be stabilized within this range. At any rate, it means that more of everything must be produced for more people.

If there is a dark cloud hanging over future abundance for all, it is the prospect of the doubling and redoubling of the world population. This book does not submit graphs, charts and columns of numbers. There are plenty of these statistics published in hundreds of books on population, available food and raw material. There are few who might read this book who have not been exposed to the frightening fact of the population explosion. It is my job to tell you about some things that each of us can do to change the impact of this explosion in the hope that by doing a million small things by millions of us in econergy, we can build a degree of comfort and safety for our own affluent lives and, also, for those who have so little.

It is not known what course will be followed by governments that will be deprived of fossil fuels and other raw materials vital to their economy and defense. In the past when access to the goods of commerce were denied a country, the result was war. With a world so well prepared for this activity, it is not hard to imagine that war will still be used as a method of getting what nations want.

These wars happen daily and are taken as the common way of life today. The world observes them on television with the same concern as they do the local ball game. Such wars are a danger to the existence of civilization and perhaps to life itself on earth. But there may be a conflict building that is far more serious and that may be much harder to contain.

We are seeing more and more conflict in small geographic areas among people who have always lived together. In the

1960's it was the violence in the cities of America and the bloodshed on college campuses, the conflict of the hippie and the street people with the establishment.

There is the plight of the financial district in the big core city, where the establishment uses the massive new buildings by day but dares not walk on the streets there at night.

It appears that as population grows it tends to spill over long-established boundaries and starts to live in stratified cultures all using the same land base. There seems to be a so-called establishment that sets the laws and collects the taxes. But there are these other cultures and sub-cultures that occupy the same area and live by a code of unwritten rules different from those of the establishment.

How often have you heard the expression, "If I did that they would put me in jail?" I wonder how true this statement is? Could it be that unconsciously each culture is beginning to recognize each other's rules and at times relax its own in favor of the other's? It is not our mission to prove this as a fact here. However, all nations are facing more heterogeneous societies and, as these differences become more pronounced, stratification is more likely to result than geographical separation. Surely the past 35 years have seen the height of geographical separation in the many new emerging countries around the world.

It is difficult to pick the next phase that will be the likely catalysts toward stratification. Race, color, religion, caste, family or tribal heritage; economic status, historical status, political concept, greed, psychological status, etc., conjure up a study of lifetime magnitude. There is, however, one division that is more evident today than any of the above. That is the very large economic gap between those who have and those who don't have.

It is this gap that must be narrowed. Almost everyone knows this. Many write about the problem. Governments talk about massive aid. All agree that "self-help programs" are best. But, what programs? And, who will do it?

The problem is so great that it is easier to just hide one's eyes from the present and take comfort in the hope that somehow government and science will solve the dilemma before it explodes in the greatest war — not a war between nations per se but between those who have and those who have not.

But the prospect of this worldwide conflict requires that all of us do what we can to change the underlying problems of food and energy shortage. It is here that each of us tends to lose contact with the problem because few of us have given much thought to where our own food and energy comes from, much less think about being our own producers and even helping the poor and uneducated gain a share of each.

There is something each person of the western world can do. We can all make it our goal to buy the equipment, when available, to supply at least a part of our own energy — solar energy, blended motor fuels of gasoline, alcohol or methanol; generators run by wind or sun to supply our own needs; fertilizer from local plants that make it as a co-product in fermentation; products made from the chemicals of organics rather than the chemicals of fossil fuels and mines.

- As skilled workers work for those new companies who produce with the econergy concept.

- As businessmen direct our skills toward running the new companies in econergy or converting at least a part of our present production to these new products.

- As scientists direct as much research as possible toward the commercial development of econergistic endeavor.

- As promoters and entrepreneurs expand as many existing industries and build as many new entities as our time and talent will allow.

- As governments relax the tight regulations and the exclusive licenses and allow free access to raw material, transportation, land, money and people.

CHAPTER VIII

A VIEWPOINT ON MONEY, WEALTH
AND THE ECONERGY DOLLAR

*Money is the root of all evil only when it is given a role above
its function as a measure of wealth.*

You don't want money. What you want are other things
for which money can be traded. So, what is it and where does it
come from? What is the value of a dollar?

As one reads the writings of many learned economists, it
becomes evident that they look upon money as something to be
manipulated by changing its value, its availability or its rental
price. This is often called "fine tuning" the economy.

It is assumed that since this manipulation is done by the
people's government that it is always in the people's interest.
The fact is that such manmade instability reduces incentive to
save — thus less financial security for individuals and less funding
for business and jobs.

Thinking about such things forces one to search for a
personal economic point of view that makes sense in a most basic
and simple form. Few of us are economists and the average business person today need not be one. It is necessary, however, to
believe in some principles that work in a world of changing
money values, changing political forms and changing economic
thought. Without an even line of thought that is basic, one finds

himself off balance and continually out of sync because of the changes. There are also as many economic theories as there are economists and it is difficult to decide whose disciple to be.

There will always be changes in political systems and economic theory. Our only hope is that in some way money value can be stabilized.

The first thing, then, is to get a personal understanding of what money is. It seems so important and we all seem to want it above all else. Why is this? It is only a medium of exchange and this means that we can exchange it for things we want. So why don't we express our desires for wealth in terms of the actual things we want?

What we are saying is that money is a measure of something else. This makes money similar to inches or meters for measuring things. How long has it been since you heard anyone say they were saving inches? This form of measure is only a measure, too, but represents a known length.

Although there is a difference between dollars and inches, the prime purpose of both is to measure. One difference between the two is the fact that a dollar is of changing value. It changes in its worth from day to day and most of the change is arbitrarily manmade. It is a measure over which a small businessman has no control.

Most people who have attempted to build their long-term security in terms of current money have been disappointed. They have found that the "ever-shrinking nature" of money has made the wealth that it is supposed to measure a better value than the actual money.

THAT BRINGS US TO WHAT DOLLARS ARE SUPPOSED TO MEASURE — WEALTH. IT IS HERE THAT THE MYSTERY OF MONEY IS BEST EXPLAINED. WEALTH DESCRIBES THOSE THINGS THAT ARE PRODUCED EACH YEAR IN THE FORM OF PRODUCTS FROM MINES, PETROLEUM, AGRICULTURAL PRODUCTION, SOLAR ENERGY AND ITS DERIVATIVES AND, IN GENERAL, THOSE THINGS THAT APPEAR NEW IN COMMERCE EACH YEAR, TO BE PROCESSED AND MULTIPLIED IN VALUE AND TO

BE CONSUMED AND GONE AT SOME FUTURE TIME.

THE DOLLAR, THEN, MEASURES THIS NEW PRO-
DUCTION AND ENDS UP BEING THE METHOD BY WHICH
TRADES ARE MADE AND ACCOUNTING IS DESCRIBED.
IN GOOD TIMES AND BAD TIMES, LARGE QUANTITIES OF
BASIC WEALTH ARE PRODUCED, MULTIPLIED IN VALUE
THROUGH PROCESSING AND DISTRIBUTION, AND THE
TOTAL RESULT OF THE GOODS AND SERVICES RE-
QUIRED TO DO THIS IS REFERRED TO AS GROSS NA-
TIONAL PRODUCT. TOO MUCH OR TOO LITTLE MAKES
THE CASE FOR THE LAW OF SUPPLY AND DEMAND.

Several books should be written to expand the last two
paragraphs into a total economic theory for econergy. Not many
would care to stand on an economic theory that could be stated
in such a condensed statement. However, this states the basis for
an economic point of view for econergy business thinking.

For an example, processing and delivery indicate a type of
multiplication of value. Take the value of air. It is generally
assumed that air is so plentiful and available that it is free to
everyone. However, when this free air is processed into cool air
in the summer or warm air in the winter and delivered to home
or office, it suddenly has a "price." It is this feature that makes
the use of the sun an attractive business for those who build and
sell equipment so that consumers can use it with only the small
costs of amortization, maintenance and operation.

Wealth, then, appears to be a fleeting thing since after it
appears, it is consumed and eventually gone.

Your home, your car, your city has an end to its use.
Even land is not forever. Geology shows us that what once was
ocean floor is now dry land; what once was glacier is now being
farmed; what once was wooded and productive is now dry desert.

Real wealth is held by those who hold the means of producing, processing and distributing. They are in the cycle of
bringing new wealth into being each year. They are not fooled
by the thought that a hoard of gold or a cache of diamonds in a
safe deposit box is real wealth. Those things have their value but
they do not produce anything and people in the future may

change their desire to hold them as things of value.

Therefore, those who would generate wealth in the future must do it from sources that will be available. Those sources are the renewables and the methods for their use will be developed by those with the econergy concept.

Saying these things leads us to the problem of doing something to establish a dollar with the same value year after year. It is obvious that this problem can only be solved by federal government, but those who are econergists can support an administration that would establish what could be described as an ECONERGY DOLLAR. Thus we indicate blending basic renewable wealth into the measure that is the dollar.

Those who would base money on a gold or a silver standard state that these old standards gave money a "real" value. In a sense they are right, not that gold and silver are money, but from the fact that they are commodities and as such, have a market value which is related to other commodity and service values. This is true in a free market operating under the law of supply and demand without control of natural or manmade monopolies.

However, gold or silver as the only base for true money value may be self-defeating. Historical misunderstanding of what money is and human nature sparks the defeat.

The misunderstanding is evidenced by those who hoard, and hoarding must be separated in our thinking from saving. For example, a person who fills his safe with gold is hoarding. The government that fills its vaults with gold is also hoarding. Neither is allowing the commodity to enter its uses in industry. Therefore, nothing is being produced — the cycle of producing and using is shorted out.

Real savings are funds deposited in savings institutions or those invested in business ventures. There funds fuel the multiplying effect of using the basic commodities to produce the gross national product.

It is evident that this discussion is leading up to some type of a commodity dollar and, you say, "This idea has been around for a long time."

66

The answer is "yes, except that the econergy dollar is based on basic commodities produced within the country in quantity and on renewable basic resources that are a major part of the basic commodity dollar base."

Funk and Wagnalls Standard College Dictionary of 1966 describes *Commodity Money* as the currency of a suggested financial system, the unit of which, the commodity dollar, has a gold value to be determined at regular intervals by an index number based on the market prices of certain commodities.

Again we see a common view of money as something to be "adjusted" as prices change. It is the effort to make the economy fit the dollar instead of making the dollar fit (or measure) the economy.

One chapter in a book of this type cannot relate the total complexity of a complete economic system. However, a strong basis for money can be presented and econergists and economists can develop the concept for presentation to federal government at some future time.

Is it possible to develop a currency that will not be subject to inflation or deflation? Perhaps not but there appears to be a way to increase the stability of money.

Assume ten basic commodities that from their raw state and from their first point of storage start the major part of the economy.

Oil	
Coal	Not renewable
Natural Gas	
Corn	
Wheat	
Oats	
Barley	Renewable
Rice	
Soy-Beans	
Raw Forest Products	

Each of these commodities can be burned leaving a small but differing amount of mineral ash. The portion that burns is the energy content — this can be measured.

Therefore, instead of relating bushels of wheat to cubic feet of gas and tons of coal and barrels of oil, which is the old problem of comparing oranges and applies, each should be compared to the most basic content of each.

Jules, Ergs, BTU's, Watts, Kilowatts, Calories are all expressions of energy units. Each can be related to the other. However, for the purpose of this discussion we will use British Thermal Units (BTU) and Kilowatts (KW). (3419 BTU = 1 KW)

Again, as an arbitrary figure, let us use as a unit 100,000 BTU's of each commodity, which for the ten commodities create one million BTU's. Then arbitrarily state that this is the value of ten dollars on a given starting date and this will be the value from then onward.

All other commodities, products, wages and services, on starting date, will then be indexed to this unit of wealth measurement.

The law of supply and demand will allow prices of everything to seek their own levels. The original ten commodities of the dollar index will also have various price differences as one becomes scarce while others become more plentiful.

This, in no way, presents a method of Price Control. Prices seek their own market level. What is done is to give the measure called the dollar a definite, continuing size. In this case, it is that ten dollars is one million British Thermal Units or 292 Kilowatts of ten selected basic commodities. (1,000,000 ÷ 3419 = 292 KW)

By definition energy is the power to do work and 292 KW has the capacity to do the same amount of work next year as it did last year. Therefore, our elusive dollar now has a definite size with which the academic economist or the most common person can work and plan. You can also ask yourself whether or not you are really worth 29,200 Kilowatts for each one thousand dollars you are paid.

The values to everyone are great in this sort of thinking. It is presented here as a project for econergists to consider, to refine, to complete and to implement.

With this ECONERGY DOLLAR, you, perhaps, will want money. You can save it, invest it and use it several years from now because you know that you can trade it for the future things you want at a known value basis.

This book deals with the energy and products that can be produced from renewable resources. The concern is with "how to get it done" rather than "how to do it." Therefore, we are dealing with the business things to do to solve a major crisis.

It is not possible to consider the energy crisis and the business things to do about it without giving high priority to the present crisis in the value of money. Not to solve this problem may negate all other efforts to rebuild the national business base.

Therefore, this brief chapter has a place in this book and in your thought and activity regardless of your job or your profession.

Never before has an issue on the monetary system been developed and promoted to government by a citizen movement. But never before in this country has the average citizen been better informed or better able to understand. Never before has the issue of sound money been more urgent.

The following ideas are presented to expand understanding and to prompt debate which can lead to refinement and implementation of the Econergy Dollar.

●1. Energy is the basic arrangement of everything. It is the positive and the negative reacting and existing in the matrix of nothingness. The search to understand this matrix is not complete. The implied existence of nothing does not seem possible — so, in today's world, we must live with that which can be measured and understood.

●2. The current money of the industrial world has no basic value except that promised by politicians. The result has been a steady shrink of money value caused by false deposits of money (wealth that never existed) to the government account.

This has been the most insidious method to steal from savers (those who store wealth) that was ever devised. The problem is one of a non-productive bureaucracy that starts to dig a non-productive hole in the economy. Since the hole in the economy can be dug to any size, or, in other words, the fiat money can be issued in any amount, the hole eventually becomes so big that it caves in upon itself. However, if the money was redeemable for a real economic unit on demand, the issue of fiat money would become glaringly evident and would result in political suicide for those responsible.

●3. Average working people have considered the understanding of money to be of interest to others but not to themselves. Perhaps this has been why it is so easy for government to steal from all of us. So look at the Econergy Dollar from the point of view of the working man or woman.

You expend your energy at work and you are paid for that energy expended with Dollars. You spend some of your expended energy, now in the form of Dollars, for current things you need and want. You also save some for future things you need or want.

If you withdraw your saved Dollars in the future (your stored expended energy) and trade them for a future need or want, you should be able to do so at the same value relationship to the rest of the economy as when you earned and saved.

Energy can neither be created nor destroyed. This is a basic law of physics.

Therefore, if you put 29,200 KW into the economy as work for which you were paid One Thousand Dollars — you saved it — and when you spend it in the future, it only buys 10,000 KW energy value of the economy — it must be assumed that someone else got 19,200 KW of your work for nothing.

If energy cannot be created or destroyed, is it not true that someone put a great amount of your expended energy into his own pocket?

The law does not state that energy cannot be converted from one form to another, or from one place to another, or, unfortunately, from one pocket to another.

●4. Money can be anything that people are willing to accept as money. Ancient economies started with barter but soon found need for a more liquid form. From time to time commodities or products were used; such as silk, oils, copper, iron, beads or salt. Eventually this developed into gold and silver as the medium, then to gold and silver represented by a warehouse receipt — a form of paper money. Today all that is left is the paper and it has lost the main features which real money must have:

1) It must be *accepted* as a medium of value in exchange,

2) It must have a basic *standard* to be acceptable,

3) It must have an accepted standard value in relation to the future so that it can be *stored*(saved) for future use,

4) It must be *redeemable* to some real economic unit if it is to survive as money.

●5. The popular economist, Lester C. Thurow, in his book, *The Zero Sum Society* (Basic Books, Inc. — New York), described the annual Gross National Product (GNP) as an economic pie. That pie in its various annual sizes is divided among various income classes in percentage amounts that vary little over time. He describes the problem of making an equitable distribution of that wealth.

The point to be made here is that just as various income classes tend to hold their proportionate share of the fruits of GNP, the Dollar, also, should hold its proportionate share of the value of the economy regardless of the size of that economy.

To understand how and why the Econergy Dollar would always relate to the current economy in a meaningful way, one must only look at the relation of the Econergy Dollar *Base* to the total economy.

Following a complete study, it would be found that crude oil, coal, natural gas, corn, wheat, oats, barley, rice, soybeans and raw forest products would hold a rather precise percentage value over time to the GNP. Four decades ago, granted to be a simpler time, the common knowledge was that the National Income was always seven times the Farm Income.

The current precise percentage of GNP of the proposed base is not available at this writing but for an example, assume the following.

If the proposed Dollar base averaged "X" percent (whatever the exact percentage is) of Gross National Product from one period to another, it would be obvious that as the value of the average of the index mix rose or fell, a multiplier of "X" would always reflect GNP. This, then, would indicate that the energy value of the Dollar Base had a direct relationship to the energy value of the total economy. In other words, the Dollar would always store its proportional share of the economy as prices fluctuated.

All the features of sound money are present. Such a Dollar would be prized, thus acceptable; it would have a precise size, thus a standard; it would have future value, thus a storage unit of wealth; it would be redeemable in domestically produced energy units (commodities). This Dollar would not be only strong — it would be honest.

There may be additional desirable things to consider in solving the social and economic problems of modern industrial society, but what could do more than this basic underlying problem being solved which would:

1) Change the society from one of lusty consumers to one of prudent savers,

2) Produce new technology, new industry, new jobs, and plenty of food, clothing, shelter and energy for everyone,

3) Promote confidence in business and government.

Is this not a desirable subject to be considered by the average citizen, by the trained economist, by the business sector and by government? As Econergists, can we accept the Econergy

Dollar as one of our top priority goals? Are other goals reachable and maintainable for long without obtaining this one first?

CHAPTER IX

WHO? YOU!

There is the old story about the mice who wanted to tie a bell on the new cat, so that they might hear him and escape his hunger. The question was asked, "Who will bell the cat?" and there was no answer.

Why don't *they* do something about the energy problem? Why don't *they* tell us the facts? Why don't *they* lower prices? If *they* can go to the moon, why can't *they* solve our problems here on earth? Haven't we often referred to *them* in this manner? Of course. This is our way of referring to those nameless, impersonal, unseen giants called government and big business; and we, as "just folks," hope *they* will take care of *us* in some benevolent fashion. So, I ask another question, "Why, in this so-called free enterprise system, have we not seen a rush to organize new source energy and product companies in this crisis period?"; and the only answer seems to be that each of us is waiting on *them*.

Of course, there is no such thing as a true "free enterprise system," and the package of rights given to individuals and business is better described by the term "private enterprise system." We can't start a business without a permit, a license, evidence of ownership, or qualification in some form by one or more of the departments of local, state or federal government. But this does not eliminate individuals or small business from participating in any legal venture. The right to be a resourceful business or businessman is still a real fact in most western countries and that right includes rich and poor, large and small, and smart and ignorant.

These precious rights, though not quite free, are still available, and are in danger of disappearing through lack of use. A few rights vanish each decade because of disuse and a future generation may look back in their history hoping to find what *"force"* could have killed such a great thing.

To lose this system by default is an unpardonable error, since it was achieved by our ancestors who knew the misery of the alternate systems. To commit such an error is as deplorable as if, for example, the children of former slaves relaxed their desire for freedom, just a bit, and allowed a few of their numbers to revert back each year to the old condition.

The accelerated exponential growth in technology and in the size of the institutions that wield the power of the technology is a future threat to human freedom. It must be remembered that these institutions are not, in general, democratic nor are they benevolent. They are organized to manage, to project, and to protect their great power. Your small stake, as a small entity, may not concern them. Therefore, we should be aware of a new type of technocratic totalitarianism into which we could all be politically and economically captured and which is being justified on grounds of economic order and science worship.

We are not discussing here the super-large industrial companies alone. Government, Government Corporations, the Military, Medicine, Hospitals, Foundations, Charities, Schools, Universities and Financial Institutions are all structured into this interlocked giantism. It is to these institutions that people flow from the education mills to take part as small cogs on big wheels, never to feel neither the joy of total personal business success nor the full impact of personal business failure. But neither do these people feel the full comfort of what they were taught to prize — *Security*.

We have all seen the catastrophe of the closed factory which resulted in lost jobs for all levels of employment — workers, technicians and managers. These factories are often closed, not because of business failure, but because a conglomerate organization perhaps elected to close it for reorganizational reasons or because of some edict by government. Regardless of the reason, those who lost their jobs lost their prized *security*.

These traumatic experiences cause those affected to consider where real security may be found. Many find it in their adaptability and their mental outlook even when changing from highly skilled, academic backgrounds to menial work. Therefore, security may not necessarily be found by working for super-size industry, or super-size government, or with a high-degreed education.

This desire for "security" by association with large-size organization is, in part, a result of the education system. The system is now a super power within itself, and it has the function of turning our professionals for other giants like itself. The schools do not intend to be narrow, nor do they *intend* to produce narrow goals for their students. But how often from grade one through the highest university degrees have teachers said, "You must learn this and further your education so that you can get a better *job*." This almost subliminal idea projection, however unintentionally, builds a narrow goal and produces a more precise size and shape to one's mind and skill so that he or she might fit as a cog on the big wheel. It is also important to the computer that people come in precise packages; otherwise the computer cannot function.

"Innovation," "originality," and "tomorrow" are traits and conditions difficult for the computer to define because they reflect change and adjustment. However, these are the tools of the entrepreneur.

The education system should be, and is, the entrepreneur's best springboard to his career because schools, like libraries, are depositories of what *has* been and what *has* been known. As such, it is the place to gain experience in its most concentrated form. With this learned experience, the entrepreneural endeavor should be better guided in its innovation and originality and in its applications to the unknown tomorrow. With this in mind, it may be better to reduce the emphasis on *jobs* and to point out to students the fact that schools are places to gain broad, learned experiences for tomorrow's enjoyment and well being.

Therefore, in this social and economic system of the western world, it is evident that satisfaction, safety, security and happiness are not always present as a result of large scale institutions. We see great misery on the part of those who lose their jobs when technology passes them by and this misery is heaped upon them by the very system for which they have been educated to serve as workers, professionals and managers.

Is it any wonder, then, that those of lesser training and education are considered unemployable in the broader world system? Can we not begin to realize how difficult it must be for that untrained individual in one of the poorer emerging countries? Is there no way to bring the businesses and the jobs down to a level that will allow participation by everyone? Are the new technologies now being developed to replace the fossil fuel economy going to be so complex as to eliminate participation by a majority of the world population? Certainly the fruits of the present system that grew from the technology explosion and fossil fuel bypassed about two-thirds of the people of earth.

We should look at a few of these new high technology energy systems now being proposed to replace the dwindling supply of fossil fuel. It is glaringly evident that I have not discussed nuclear energy in this book. It is not that I disapprove of it as energy but rather that its use is only practical in the most advanced societies and by the largest business institutions. As an econergist, I must bypass this as an endeavor in which I cannot personally participate. Do I wish that it would go away as a method to produce power? No! Let's see its use grow and especially the technology grow. But slow down! Are we building production or destruction? Do we really know how to dispose of obsolete plants and atomic waste? Are our political institutions and human nature more tuned to energy or to the kind of power gained with bombs? Can the present "have nots" use this new force? Can we afford new systems that do not include everyone in its benefits? Need I say more about nuclear energy? It is a question in my mind, so I ask only questions.

Then, there is the proposal before the technical community which would build large satellite bases in space in which solar energy would be collected and beamed to earth for use as electricity. There are proposals for very large space communities manned by space station workers who would keep the space power plant working. Costs of this will be massive. Can this really be done? I can only believe that it can but that its accomplishment would benefit very few at the expense of many. But should this sort of thing be considered at all? And, again, as an econergist, I must say "yes." One important thing which must be done is to keep such projects in proper perspective. Let us not drain a greater part of research and investment funds from the many simple, small things that most people can use *now* with present technology.

There is a technology so simple that almost all levels of competency could participate at some level. So is it not possible, for example, to make the words *small, simple, meaningful, personal, community, direct, excellent, sufficient, renewable* and *econergy* replace the words and phrases of the past business cycle — *super, multinational, military-industrial, complex, technocratic management, cartel, monopoly, exclusive license by government, oil* and *controlled*?

After talking about this sort of thing, many people see nothing but "bad" in our present bigness. They see a need to do away with this big size by killing it. But wait! This is how the world *is*. This is who we *are*. This is the life blood of today's commerce. It is what each of us helped build upon a foundation of finite fossil fuel. Right or wrong, it is the *now*. The trick will be to change the foundation *gradually* so as not to collapse the total social structure built upon it.

Everyone has opinions on the subject and some have very learned ones. For example, a current popular book, *Small Is Beautiful*, by E. F. Schumacher (Perennial Library — Harper and Row, Publishers) has caught the fancy of many readers around the world. His discussion of the problem is from the viewpoint of a trained English economist. But his solution to the problem is a paradox.

He sees smallness in the sense of one's operating under the umbrella of socialized government ownership. He sees private ownership as immaterial. He sees the way to handle the present big corporations to be the free issuance of stock, both common and preferred, to government in an equal amount to that which is held by investors.

It is my opinion that such expostulations are made by one who has never been "in business" and knows little of what motivates business and business people. Such viewpoints play upon what may be the dangers of today's bigness, but replaces it with making already big government omnipotent. The result would not only be more bigness, but also disaster.

As econergists, it is hoped that you and I look at how things are *now*, and use these things to gain our goals by *blending*. Large institutions have many things we can use — people, technology, money, patents and *problems*. In spite of their size, they need small business.

Take, for example, the need to utilize the hot water from coal-fired electric plants described earlier. If you have a greenhouse idea or another use for hot water, you may find yourself a welcome guest at the highest corporate level of the big utility company. Also, many of the largest chemical companies have patents to utilize organics, just waiting for some small manufacturer to use them for a small royalty cost. Government has thousands of pamphlets and booklets on alternate energy to be used by *all* citizens. Universities have information and people to help you do about any technical thing. So, our big institutions are there are for our use in building relatively small econergistic industry. Blend what they have with what you need.

This point of view is not destructive. It does not destroy today's big institutions — especially big business. However, the next generation may find the present large multinational conglomerate hard-pressed to compete with the relatively small econergistic enterprise. If it becomes too difficult to compete, the big energy company or the big manufacturer may change what they do. They may wish to use their big money pool more as bankers rather than as producers or marketers. The future may see them change.

Whoever said that things would remain the same or that what is today's problem could not change into tomorrow's asset? the answer to that which could be lies with *you*, not *them* — and surely many of you are saying, "Who, me?"

I must look at just *you*. You may be working at almost any job from a trash hauler to an executive in a large corporation. You may be a small businessman who would like to expand into product manufacturing. You may be a housewife with a great desire to have a business. You may be an engineer, a chemist, a janitor, a salesman, a doctor, lawyer or Indian Chief. You may be rich or poor, black or white, young or old. All of you are candidates to be econergists in name and especially econergists of action. Most of you don't have the money to start your own business and most of you think the rich man is all who can or will do this new thing.

This thinking must be changed. You must realize that the rich man has already taken the plunge and has already made a success of his business. He is quite busy with his many endeavors and may not be the best prospect to promote a new venture.

It is the average person who desires greater affluence who must start new things. He has not "made it" yet but believes that he would succeed if he could find the right business. When he does find the right business, he seldom has the money to be the sole owner.

Therefore, you must believe in the system that made the "big" companies "big" and follow their example to the size which your project justifies. Many methods of pooling money exist and many individuals are looking for reasons to invest if the features of safety, growth, profits and liquidity can be presented in a way that reflect a reasonable risk. These people do not live in a far-off city; they are your neighbors. They, too, are seldom rich but would like to invest in a local industry that would grow and prosper. These people want to win, as does the entrepreneural group, and should be treated with the same regard as the most ingenious management. The money, therefore, is right outside your door, regardless of where you live.

As indicated earlier, many of the new econergistic industries need a period of engineering and testing whereby known techniques are arranged into new sequences that will result in new energy-producing equipment. Some of these will result in patents and some will not. Much of the work that you think needs to be done may have already been done by others. So look into it. Don't try to be secretive in your approach. Learn to exchange ideas with others in this new business. There is room for all and the success of your competitor may be important to your own success.

Keep in mind a goal of excellence and service. These are difficult goals to fully achieve but they are more likely in small organizations. Average companies never grow and will disappear or be acquired by those who do excel.

And, with that said, we begin to see large companies building again through growth and acquisition. This is normal and healthy until the size reaches the point where companies become larger than most governments. When companies become so large that they have their own international security forces, become involved in international intrigue and covert activities, and when these actions sway international government decision, they are too big. When companies control any domestic industry, they are too big. But a one hundred million dollar investment, for example, is not too big if that is required to produce a given product. Therefore, size is acceptable relative to the job to be done rather than to the amount of power to be wielded.

Again, most beginning econergy companies need not have more initial capital investment than a family-sized farm. This size surely includes *YOU*!

So the purpose is to present a concept upon which *you* can organize or expand a business structure to produce the new equipment for the new age of ECONERGY. The field is wide open, the need is great, and you are the most important part required to make the concept operable.

WHO? YOU!

PART TWO

PROOF - TO CHANGE AN IDEA

TO A BUSINESS REALITY

A book on concept calls for a demonstration.
If another cannot be found to do it,
candidates for the job narrows to one — yourself.

CHAPTER X

THE ECONERGIST

Individuals of all types being allowed to cut through the maze of endless impossibilities on the narrow road to the possible has been called the American Way. Econergy widens the road of the possible and, in the true American sense, invites each to follow its course.

We have discussed a basic concept that is concerned with producing wealth from a place best described as "Nowhere and yet, everywhere." At least this wealth is from a source that we can use without depleting anything. What we are really doing is using the existing energy around us and changing it from one form to another or from one place to another.

This book has tried to point out that a whole new level of business is waiting on you and me to "get in" and participate. The book presents concepts and examples of how to think about the new horizons of business to be known as *ECONERGY*. The word is not *"Techergy"* which would stress the technical things to be done. It is *ECON-ERGY* which emphasizes the business things to be done.

All of the seminars, meetings and conventions on alternate sources of energy spend their full time on the technology of the subject and little on the subject of how to organize the known Technology into a business reality. The reason for this is obvious since the average person, together with the technician, is intrigued with "how it works" instead of with "how to get it in production."

We can go to energy meetings all around the world almost any day. We can hear of the marvelous new things at each meeting. We will meet researchers and engineers who are anxious for someone to organize a business which they could join. We go away from the meetings with a broader knowledge of the new technology and new proposed research but at next year's meeting, we will have seen few, if any, commercial applications.

Therefore, *ECONERGY — THE CONCEPT OF PLENTY* is concerned with getting the job done.

We further discussed the good economic impact of a new industry boom around the world. We stressed proper use and proper husbandry as important to the environment and to the success of the concept.

A brief overview was presented on a few types of new industry that fit the econergy concept. Not much in the way of detailed technology was given because it is so available in about all libraries and schools. The average consumer only *THINKS* he wants to read the technical descriptions of how such things as solar furnaces work. It is my opinion that he wants to know that it does work and that his next desire is to know "where to buy it" and "what is the price?" So, as promised, this book is not technical. It is a book that, hopefully, presents concept in enough depth to give you a springboard into your own econergy business or, at least, give you a reason to join such a company in the capacity of your choice.

We even discussed some of the ways whereby new industry could be promoted and where and how the new money could be obtained. We considered that you might become your own promoter and build that new small business to make a piece of equipment to generate a new type of energy. We suggested that you be in the business of selling individual energy—generating equipment instead of producing and selling energy.

There were not a lot of graphs and tables of numbers to point out the problems of food and energy production and the population explosion. There was a discussion of a positive concept of what you and I can do to meet the challenge of these problems.

This brings us back to just you and me. As you have read, you either agree or disagree with what was presented. If you disagree, perhaps you have another approach to the problem. More power to you. If you agree, however, you are now an *ECONERGIST*.

To be an econergist is not more difficult than to be an environmentalist, ecologist or conservationist. In fact, it is a practical thing to be in addition to one of these. All three of the above disciplines are hard-pressed for positive things to do. The ECONERGIST fits in well here with a practical plan which allows caring people to protect the environment with an economic plan that will support the social and economic structure.

So, as the case is made for you and me as econergists, we may have a kinship of thought and purpose. At this time of writing, it is difficult to know where this kinship will lead us. Could it be to work together in a new company venture in econergy? Could it be in a formal econergy club? Or could it be that you and I just understand a great need and that we support each other in attainment of our separate goals?

Regardless of where the concept leads us, one thing is sure. A concept. like a new invention, is of little value if it cannot be demonstrated.

It was in this area of demonstration that another two years was to be added to the travel and writing in which two years had already been spent.

CHAPTER XI

THE CREATION PHASE
(To Choose A Project)

It is not a brilliant mind but a receptive one most likely to be creative.

As one reads the concept of econergy and begins to feel the need for more concrete examples, it becomes evident that it is difficult to find demonstrable situations that can be fully studied. Many businesses may consider the entrepreneural effort that established their enterprise as confidential. Others may have lost the record of the real things that happened in their initial stages or the present management team may not be the original promoter group. Therefore, it was necessary for me to start a new project myself.

The guidelines under which I would choose a project were defined and, in general, they included the following:

THE PROJECT —

1. Would be simple.
2. Could be duplicated by existing companies.
3. Would require minor investment in tools or equipment.
4. Could have a national impact.
5. Could be accomplished with meager technical expertise.

6. Could be practical in both the industrial nations and the third world nations.
7. Could be promoted without major personal cash investment.
8. Could be done by private industry without government grants.
9. Would help prove the possibility of being innovative in today's industrial world.
10. Would lend credibility to the meaning of econergy.

With this list of desirables to guide me, I fully expected to promote a business that would convert organics to chemicals or burnable fuels. This was an area in which I had previous experience. I thought about many bio-mass projects and I probed several possibilities.

Alcohol as a fuel was exciting. Methane gas was exciting. Char and tar products were exciting. Several greenhouse ideas were exciting. But none had the features to apply as an example for use by a large segment of existing industry. Few could fit inside the ten point guidelines.

The answer came when I asked myself the question, "What would people want most from an econergy-type business?" First was the desire to have the home and work place at a comfortable temperature at a reasonable price. The rising cost of being comfortable was reported daily in the news media. It was not difficult to decide that this was to be the area of my new demonstration project.

I made the decision to concentrate on a simple method that would keep a building in the comfort zone of 70°F to 75°F.

Using the sun to do this in winter is a simple thing. But I was now in Phoenix, Arizona, and the heavy power load is not for heating but for summer air conditioning. If the method was to work for both heating and cooling, and be simple, I would have to think about it for a while.

Again the answer came, not from a new invention or a new technology but from a local ancient art. The first settlers of the productive Salt River Valley, in which Phoenix is located, were not the early Spaniards or other Europeans, but the Indians. These Indians were farmers. They irrigated the hot, dry, desert floor from well-engineered canals supplied from the Salt River. Today's extensive irrigation system uses the original canal routes laid out centuries ago by this Indian society.

The questions now were evident. How did these Indian farmers live? How did they cope with 115°F summer temperatures and 30°F winter nights? Were their bodies more adaptable to temperature extremes than ours? Could it be that an ancient culture capable of engineering canals to modern specification standards would have something to tell us through the ancient ruins of their homes?

There are ruins on the desert floor close to their farms that were built of massive dirt walls much like the adobe walls of the early great plains pioneer. The Spanish mission churches which are evident in the Southwest U.S.A. copied this type of building and most are made with very thick adobe walls.

Heat penetrates a thick earthen wall slowly. The sun shining on such a wall will cause the outer 18 inches to become hot. However, if the wall is 36 inches thick, the heat will not reach the inside wall surface. When the sun goes down and night radiation reverses the energy flow, the hot portion of the wall cools as energy flows away from earth to the night sky.

Also, there is a second type structure found in the canyon and cliff areas of the mountains above the valley floor. These are the cliff dwellings found in the crevices of the great stone cliffs of several Southwest areas. The most profound example of passive cooling and heating with radiation is to be noted as one looks at such structures. These were to be the example upon which I would develop a modern cooling and heating system using rocks as the most exotic solar device.

My grade school teacher taught me that these unique structures were built high in the stone face, with ladders to climb to the first floor and ladders to climb to subsequent floors as a fortress against enemies. But this makes no sense. An enemy need only camp at the base and starve the occupants out.

It is further noted that some of those who lived in these

structures also farmed and that their farms were as much as thirty miles from the cliff dwelling. This indicates an obvious fact — that the ancient Indian prized his home temperature comfort level as much as modern man; enough, in fact, to walk many miles to and from work to enjoy such comfort. There may have been strong social factors that influenced this living mode, but the temperature comfort factor had to play an important role.

The heating and cooling principle that keeps the cliff dwelling in the comfort range is ingenious and simple. It operates on the effects of solar radiation on mass for heating and the effects of night sky radiation from that mass for cooling. Therefore, it is a system that passively sits in the radiation stream and stores and releases energy as needed to maintain a comfortable temperature.

Most of the structures were built on great stone cliff formations in a south-facing crevice with a large rock overhang. In winter, the sun is low and shines under the rock overhang striking the windows and stone walls of the living areas. This heats the walls and the great stone cliff. Many million British Thermal Units (BTUs) are stored in this fashion. As the stored solar energy leaves the earth to the night sky the walls cool, but not much on the inside part of the walls. As a result there is little inside temperature fluctuation.

In summer, the sun is high in the sky. The rock overhang shades the walls of the apartments all day. There is no direct solar radiation passing into the walls. The only heat transfer is from the warm ambient air but this does not penetrate to the inside before night. What is stored from the ambient air leaves the wall at night on its way to space and the inside temperature at the living areas remains comfortable and stable.

From this ancient example came a project that can meet the ten point guideline test. It can be a business expansion for thousands of contractors. It will not be exclusive to anyone. It will fit all areas where rocks can be delivered.

I picked the lowly rock to demonstrate one simple application where it can ease the demand on fossil fuel. I, also, depend

91

on the lowly rock to be a good example of econergy in action.

Now, for the first time in this book, we will look at the state of the art of cooling and heating with rocks. Call it a technology if you like because everything that we do has its so-called technology. However, it is more like saying that we now have the technology to have hard boiled eggs. If I were to ask you what this technology was, you would say that "you put an egg in boiling water for five minutes and it comes out hard boiled." With this I would agree.

However, if all of us were required to know the amount of heat in terms of BTUs per gram of egg or the chemical changes that happen to change a messy white and yolk to the solid state, I'm afraid few of us would ever eat hard boiled eggs.

This investigation was the result of performing a search for a new project. It is the first of the four parts of a promotion — *The Creation Phase*. Next comes the Documentation Phase, then the Research and Development Phase, to be followed by the Production-Marketing Phase.

CHAPTER XII

THE DOCUMENTATION PHASE
—THE LOWLY ROCK

The lowly rock is the most exotic agent in heating and cooling buildings.

It is always a relief to finally make up your mind. It is, also, unsettling if you have made your mind up to attack a project in which you have little background experience and you note that there are few others who have experience either. I did think that I understood the principle of cooling and heating a building. But what did I know about using the lowly rock to perform such a task?

I knew that my knowledge on the subject was small. I also knew that, however small, the time had come to document what I did know and arrange these bits of knowledge into a sequence that would indicate a technically sound approach to cooling and heating a building with rocks. This had been done in various ways in the past but the most important step in promoting the idea into the reality of a commercial venture had never been made. The work done was on the technology. No one had made the effort to make a business exist using the technology.

My work in documenting the information was to be aimed at attaching a simple process to a business group that could make the project happen on a national basis. I did not know who this group would be. Therefore, I wrote the first document in the form of a memo. It was directed as much toward myself as anyone else. It was a crude document. It shows the unsophistication of the beginning of most new promotions. It did, however, zero in on one industry group of which I knew little.

Therefore, with the belief that the ancient Indians of the Southwest understood a basic principle of keeping their homes at a comfortable temperature, and with the belief that rock heatsinks could duplicate the concept of mass in individual buildings, the following edited memo was prepared.

MEMO

SUBJECT: Notes on ROCKS AS AN EXOTIC AGENT TO USE SOLAR ENERGY

BY: Robert D. Bowers — July 1977

ABSTRACT: One of the most pressing needs to utilize the sun is that for cooling and heating buildings. This indicates a need to maintain building temperatures at approximately 70-75°F. Such a temperature range indicates "low grade" energy. "High grade" energy would indicate the need for temperatures to make such things as steam to turn the wheels of industry.

Emission, transmission and absorption of solar energy describe features of "radiation." Radiant energy arrives in waves of heat and light as it approaches earth, and such energy leaves the earth as radiation to the deep night sky. This indicates two forms of energy transfer that can be used:
(1) Energy arriving during daytime when the sun is shining;
(2) Energy that leaves the earth as it radiates to the deep night sky.

Therefore, there are two modes of sunshine. One that is present during the daylight hours and one that is absent during the nightime hours. These facts allow the use of low-grade energy to be used in radiant form for both cooling and heating at the above-indicated temperatures.

RESEARCH: Most of the work being done on the use of solar energy consists of attempts to research and re-research known facts for the purpose of converting the sun's radiation from its present energy form to another form so that it can be used to power fossil fuel type appliances and engines. It is this changing of energy form which has not been economically mastered at this time. However, the use of the sun's energy through its properties of radiation does not require a change in energy form — only a transfer from one body to another. With addition of mass, the feature of storage is available. This simple and practical application is demonstrated in the rockbed storage system and allows immediate use of our present state of unsophistication in the helio-sciences.

Such a system is passive except for the addition of an air moving fan. Obsolescence of impracticality will not occur by the introduction of some yet unknown exotic discovery. This system is so basic that it is the ultimate in exotics; it is, also, so old in the art that it has been overlooked by many. It is the system that was used by the Southwest Indian in his cliff dwelling mode.

Rockbed storage allows a building to be self-contained in its heating and cooling requirements at little cost over conventional construction and with the feature of greatly reduced power bills.

DESCRIPTION OF PROJECT:

The rockbed storage system of home and commercial construction is one that utilizes the passive capability of rock, concrete, insulation and air to be combined with the capability of a small fan to furnish the dual objects of total heating and cooling. The system is made up of the following components:

(1) A solar air collector of almost any form. (A vertical south-facing glass wall known as a south wall collector; a roof mounted solar air collector of almost any standard manufacturer; or, for desert regions, a simple metal roof painted a dark color.)

(2) A rockbed storage unit sealed in a concrete vault or a plastic film-lined pit or other material with substantial air inlet and substantial air exit forced by an electric fan.

(3) A "swamp" cooler or a small refrigerator-type air conditioner and auxiliary heaters in cold climates.

(4) Air pipes from the collector to the rockbed storage system for wintertime use. Air pipes to the refrigeration unit, designed so that air can be drawn directly from the cool night air. Standard air piping system to the various rooms in the building.

For winter use, the system is integrated with air pipes so that air circulates from the collector to the rockbed storage, through the building, and returns to the collector. For summer use the collector is bypassed and the rockbed is charged by cool night air or when necessary by night air cooled further by an air conditioner.

This feature of heating and cooling by radiation extends the function of the sun's energy to the building system without changing energy forms. The only limits to extended use during prolonged overcast and cold or exceedingly hot nights is in the amount of mass in the rockbed storage. Therefore, the system is a 100% system, being only limited by the practical economics of how much rockbed storage is installed. If supplementary heating or cooling is needed, such appliances can be operated in off-peak power periods to recharge the storage system with cool or hot air at off-peak rates. (These rates do not exist in all areas but are under consideration.)

It must be assumed that there are factors that will limit the size of the rockbed storage, which in turn will make additions of cooling or heating units necessary, depending upon whether extremes are toward hot or cold, or both, in a given area. So-called "normal" weather can be handled almost anyplace. But suppose that conditions of cold or hot in any climate are described as (1) normal, (2) severe, (3) extreme. The following may be the order of operating the system:

(1) *Normal* — Operate the system in its natural state either heating or cooling the rockbed storage with only outside air, or air from the collector.

(2) *Severe* — Operate the auxiliary heating or cooling units to recharge the rockbed storage.

(3) *Extreme* — Use auxiliary heater or cooler to pump hot or cold air into the building directly. (The more practical application may be to use the auxiliary units in such a manner that they always push their warmed or cooled air through the rockbed storage prior to entering the house or building.)

97

The rockbed storage system can be used with almost any type of architecture and most types of commercial solar air collectors. The one that would work in many areas, and especially in the southwestern deserts, would be a standard, well-insulated house with a metal roof. This roof would act as a collector and the temperatures at the ridge pole on an average winter day may range from $90^{\circ}F$ to $120^{\circ}F$ — depending upon color, orientation, size, pitch, ambient air temperature and wind.

In areas where efficient collectors are needed, the south wall collector on the south wall of a building will increase solar heat collection. Glazing the metal roof will upgrade efficiency. These systems are pleasing to the eye and require very low-cost materials that are available in almost every community. All can be built on site with simple home construction techniques.

MISCELLANEOUS INFORMATION:

1. It requires one BTU of heat to raise a pound of water one degree Fahrenheit. This means that the pound of water has stored one BTU of heat. Therefore, a ton of water (2,000 pounds) will store 2,000 BTU's of energy for each one degree rise in temperature. The average rock will store about .20 BTU's per pound per degree Fahrenheit. Therefore, it will take five times the tonnage of rocks to do the same work as one ton of water in heating or cooling, but the costs of installation and the problems of handling water make a massive rockbed storage system more practical from a cost and operational point of

view.

2. *Heating Example* — If we assume a 1200 square foot house, with a rockbed storage of the same square footage and seven foot deep, there would be 8400 cubic feet of rockbed storage. If the usable rock weighed 75 pounds per cubic foot, there would be 630,000 pounds of rock in the storage vault. Since rocks store approximately .20 BTU's per pound per degree Fahreheit, rockbed storage would be storing 126,000 BTU's of total heat per degree Fahrenheit rise. If the rockbed storage peakedat 100°F each day, and the house temperature was to be kept at 70°F, the total BTU's available would be 126,000 x 30°, or 3,780,000 BTU's available to heat the house. If the outside temperature was such that the house lost 500,000 BTU's per day, the storage system would have a reserve of 7.56 days, without any sun on the collector. This reserve would be reduced as the temperature got colder and would be extended on milder outside temperatures. (Author's Note: The basic logic is correct. However, a weight of 93 pounds per cubic foot of rock would have been more accurate for the rocks of the Salt River.)

(Excellent house insulation is a must for any heating or cooling system.) (1200 x 7 x 75 x .20 x 30 = 3,780,000)

3. *Cooling* — It will require about 15,000 BTU's of energy for each degree day of cooling or heating for the average house in the United States. Well-insulated homes will require less. Therefore, the cooling of

99

buildings in the warmer climates may require rockbed storage units as large as the more northern climates will require in heating units. Properties of heat are such that the heat of a warmer body radiates to the colder body. In other words, the rockbed storage is sucking the heat out of the house and storing it as the air moves through the rockbed storage unit. The theory then is the same for cooling as it is for heating. The only difference is that the system is reversed. That is, in the heating mode, the rocks are giving up their heat to the cooler air from the house in wintertime, and they are accepting heat from the warmer air of the house during the summertime.

4. *Operating Areas* — A rockbed storage system is an applicable system to both northern and southern climates, and to both extremely cold and extremely warm. In climates such as Flagstaff, Arizona; Denver, Colorado and Salt Lake City, Utah, the system may very well be a 100% system requiring little or nothing in the way of supplementary heating or cooling equipment. The availability of sun during prolonged cold spells in these areas will keep the rockbed storage charged with warmth. The cool summer nights in these areas will recharge the rockbed storage system in the summertime without need for auxiliary cooling or refrigeration units. More extreme climates such as Buffalo, New York; Minneapolis, Minnesota and Seattle, Washington (overcast), will require auxiliary heating units to supplement the rockbed storage during extreme weather conditions.

In areas such as Phoenix, Arizona; El Paso, Texas and Palm Springs, California, the extremes do not run toward a cold winter but rather to an extremely hot summer. In these areas, no supplementary heating unit need be considered but cooling equipment may be required as part of the cooling cycle. The efficiency of using an air conditioner results from cooling the already cooled night air and storing it in the rocks.

5. *Acceptance by the Public* — There has been little acceptance by the public for solar homes. The main reason for this is the fact that they are not readily available. Where they have been made available, the costs have been substantially higher than conventional housing. Much of what has been available has involved a lot of expensive hardware that usually sets on the roof of a house and such hardware has not lent itself to acceptable aesthetics on the part of homeowners. If opinion counts for anything, I would say that the buying public is ready for a new energy-producing home that works and does not stray too far from old established eye-appeal formulas. The average public has been led to believe that the utilization of solar energy is such an extremely high technology business that it would be impossible to understand the workings of the mechanisms. The selling job will be one of letting him in on the secret that was known by the cliff dwelling Indians and even by the primitive cave dwellers of the ancient past.

MARKETING AND SALES:

Marketing is a function that can be considered as that of organizing a *system* under which *sales* can be readily made. The sales function, then, is to use the marketing system as a vehicle from which the customer is confronted with the project or product; and, because the industry, the customer and the salesman are properly connected to the marketing system, a sale is consummated. It appears possible that a marketing system can be built upon the idea of the lowly rock as an exotic solar agent. How it works should be the function of a "show and tell" program that is coordinated through well-engineered show homes, by an informed sales approach, through a properly organized public relations program by the sand and rock industry at the local level. The cement industry is naturally involved with such a development and should participate. However, the increased use of concrete will naturally follow the increased use of rocks, so the sales effort should be kept at the level of the local sand and rock company and the local ready-mix concrete plant where the home builder and home buyer can relate on a direct basis. A proper approach would appear to be that one rock company start a project by building a rockbed storage solar home. As the engineering and construction on the home progresses, cement manufacturers could have a proper function in coordinating demonstrations and seminars for others in the sand and rock and ready-mix concrete business. The flow of knowledge would be from one sand and rock or ready-mix company to another and from each of them to the local contractor and consumer. This is an idea which could be developed into a marketing system with a national and international impact in the use of an alternate energy source. Since the business entities and the products are in place at the present time, and since additional capital formations are not required, the results to the construction business and the energy business would be realized quickly. This project could set a good grass roots example for simple solutions to present energy problems. Its impact in the heliothermal section of sun use may well spark other industry sectors to find simple solutions in the helio-electric and heliochemical areas.

Note: The above memo started the thinking and guided the actions toward the next phase in a promotion — The Research and Development Phase. It was in this next phase that simplification and refinement was to happen.

CHAPTER XIII

THE RESEARCH AND DEVELOPMENT PHASE
—To Prove The Point

A receptive mind, a desire to prove the point, and a willingness to yield to another's need leads a promotion down an unusual path.

The time had arrived to prove the point that rock bed cooling and heating was a worthy project for a local sand and gravel company or a local ready-mix company to promote to local contractors. This effort must be done by an ethical group because, if successful, it will result in a business expansion for thousands of contractors and great reduction on fossil fuel demand. With this significance, my job now was to find such a company.

The telephone book listed the companies, a call to a banker confirmed the financial strength of the major producers, and a call to several home builders confirmed that their subcontractors used these major companies with satisfactory results.

I was a new short-term resident of the Phoenix area. I had no close buiness contacts upon which I could rely for counsel. The selection would be made more from a hunch than from reasoned facts.

A drive past several operations presented all to be busy and business-like. For some reason one seemed right. I would choose

Arizona Sand and Rock Company.

The time was July 1977. The business climate was looking better. Phoenix was recovering from a severe business slump in which the construction industry had been devastated. With the economy on the upswing, timing for the project loomed good.

As I entered the front door of the office building at Arizona Sand and Rock's main plant, I wondered how strange my presentation would sound. Whoever had presented such a plan to an industrial company such as this? But here I was with my memo, "Notes on Rocks As an Exotic Agent To Use Solar Energy."

The receptionist suggested that I talk to Mr. Burns Crossland, but he was not in the office at that time. Several attempts were made to arrange an appointment but schedules were tight and, as yet, we had not arranged a meeting.

On one occasion I stopped by the office. This time I was directed to Mrs. Donna Clark, Administrative Assistant to Mr. Elmer D. Clark, President. This was the meeting in which Arizona Sand and Rock would start the first econergy project. It was not that they had chosen me, or econergy, or my memo on rocks. It was that I had chosen them.

Donna Clark is a positive representative of her company. In the exchange of information on rocks I was, in turn, given an insight to the company. Upon completion of our visit she made the statement, "If rocks can do these things and if you need to talk to someone who knows about the rocks of the Salt River, you should talk to Mr. Clark. I will give him your memo and make an appointment."

The reason for relating this early meeting is the fact that it makes one of the most important points as one changes an idea into a reality. In developing the new and the untried, the business is about tomorrow. Tomorrow is gained one second at a time. A project moves one step at a time. You must recognize when you are winning. You don't win it all in one package. Know when you have made a gain. Accept the gain. Don't qualify it; accept it and move to the next step.

This was not the time to go shopping for another company

to see if a better deal existed. What more could any stranger expect than what was stated by Donna Clark. The first econergy project was started.

Many meetings were to follow with Elmer Clark. We discussed business in general, the energy crisis, solar energy, rocks as a heat sink and many things that related to the concept of econergy.

As the weeks passed, time became a problem. Arizona Sand and Rock was involved in the midst of the biggest building boom ever in the Phoenix metropolitan area.

It was evident that the pressure of this boom would interfere with company participation in the Rock Bed Project. They were now understaffed to manage the impact of the new business volume.

During this growth in business volume, I was retained as a consultant to prepare a study for the company. The report would be written to show how the company could use its rocks in an econergy project. It was December 1977. Completion of the study would require less than three months.

The following is the information on the front page:

AN ECONERGY PROJECT

ROCKS AND CONCRETE AS EXOTIC
SOLAR MATERIALS FOR COOLING
AND HEATING BUILDINGS IN THE
SOUTHWEST U.S.A.

THE FEASIBILITY OF IMPLEMENTATION
PREPARED FOR:
ARIZONA SAND AND ROCK COMPANY
ELMER D. CLARK, PRESIDENT

PREPARED BY:
ROBERT D. BOWERS

This was the first report ever written using econergy as the basic concept of a study. It will not be reproduced here because many of the points covered have been explained in other ways in this book. However, the Table of Contents gives a clue to the tone of the report and to the approach which I hoped the company would take.

CONTENTS

CONTENTS (continued)

The company furnished an office from which I would observe their operation and their people. The boom was heating up, business was good, and everyone was frantic with workload. There was no time for visiting or coffee breaks. There was plenty of coffee but all took it at their desks along with the business action.

With a sure hand, Elmer Clark guides a complex company made up of operations in ready-mix concrete, sand and rock products, sacked rock and cement products, all of which are supported by the mining operation which brings the sand and rock from the pits. Their rock and gravel pits are in the floodplains of the Salt and Agua Fria Rivers.

The winter of 1978 saw Murphy's Law at work. It states that anything that can happen will happen; and it did. The dry desert turned wet and a devastating flood hit the company which operates in the rivers. The miles of conveyor, the heavy equipment, trucks, and processing machines were moved to high ground. The fast flood waters ran for weeks and made months of recovery work necessary.

It was not the time to start a new econergy project. I was to stand by until the company got back into production.

In May I started to visit with Elmer Clark again. Business was no longer just good, it was a madhouse. The demand for construction materials, both gravel products and ready-mix, far outstripped the production facilities of all the suppliers in the valley.

People to work in these industries were becoming hard to find. Elmer Clark was pointing out this problem to me as I pushed to start the Rock Bed Project.

He came up with a solution. If I would come into the company and work in the sales department I could, as time permitted, develop the Rock Bed Project. I would work under Burns Crossland and call on the engineering contractors. These are the contractors that build streets, roads, underground pipelines, parking lots, railroad spurs, building pads and, in general, use a lot of rock and gravel products. These are, also, a class of contractor already equipped to build the rock bed solar storage unit.

The new idea of having a job was a shock. I had not worked for a company other than my own projects since World War II. I knew little about the products that I would represent to the contractor group. I also knew nothing about the contractors.

The arrangement for the job was that we would "play it by ear." I was handed the keys to a company car and told to call on the "A" - Licensed contractors. My first calls were to be more of a public relation type. From that I would "pick it up." So the "play it by ear" arrangement went beyond my deal with the company to include a "train myself" program out on the firing line.

It was not difficult to pick out the things to be done. The problems were many and all of them were immediately important. My job was to help the contractor use the company system so that he might get the material he needed when he needed it. There was no need to sell. Everything had a customer waiting months before it was produced.

In the midst of this business climate I started to design the rock bin by myself. I looked over my shoulder often for that engineer that I hoped would come and help. But there was none.

An hour at a time, two hours at a time, I strained my small engineering knowledge to complete a rock bed design that would be the most economical for the Southwest U.S.A.

It was time to build. Summer passed, fall passed, and it was January 1979. I was anxious to start the construction.

It was Murphy's Law time again. It rained and it rained. All construction was closed down. The flood threat was here again. And it came.

Suddenly, with rain pouring, I got word that we would start construction on the rock bed unit. We started.

I had several million dollars' worth of heavy equipment at my disposal to dig the hole and backfill the rocks. A small backhoe might have been sufficient but as long as it rained we were really well equipped.

The Research and Development Phase had finally got to the pilot plant stage.

The rock bed cooling and heating unit has three parts: (1) the rock bed, (2) the damper system, (3) the collector-radiator. We build the rock bed first, installed the fan and connected it to the air supply and air return of a dispatch building. By removing the covers on the air junction boxes the bin could be recharged with daytime air for heating or night air for cooling. We tested this portion in April and May. We were able to keep the supply air at about 65°F for cooling. In May the first 100°F days arrived. The women working in the dispatch office were wearing sweaters. It worked.

It was now time to install the dampers and air direction system to be followed by the metal box on the roof that would be the collector-radiator. We installed the damper system and set the air ducts to the roof. It then became difficult to free up plant personnel to work on the collector-radiator. We were fighting time and unprecedented material demand.

It was June. If we did not get the construction completed quickly we would not test the night sky radiation cooling that summer.

What had been a madhouse last summer was now worse. We were trying to supply an ever greater demand for product with a production complex that was still damaged from the past two floods. The rock bed project suddenly had a priority that came right after broken lead pencils on the priority scale. As summer waned, hope faded for the night sky radiation test.

With August and September gone I must admit to being discouraged. It would be summer of 1980 before we could test in the extreme summer heat. No need to push now. We would finish installing the collector-radiator and start up in February or

March, 1980.

The workload was unbelievably heavy in the gravel and rock product business and I had about all that I could say grace over. I really did not have time to give the rock bed a proper priority.

As 1979 drew to a close, we could see signs of a slowdown in the construction business. The normal government answer to adjusting the economy was beginning to work. Their answer to times that are too good is to generate bad times. Their policies, especially high interest and tight money, were taking effect. As usual, the construction industry would see the bad times first.

There would be time now to complete the rock bed project. We ordered the sensors that were needed. We finished some small items on the now completed collector-radiator. We made some initial tests on the unit to see how well it worked and we were excited with the fact that it heated the air to much higher temperatures than we had ever hoped. The time had come to get down to business. It was February 1980.

Murphy's Law could not stop us now. The mathematical probability of a third One-Hundred-Year Flood was remote. It could not happen.

The winter storms were back to back across the Pacific Ocean. The Los Angeles area was having great difficulty with floods and mudslides. But this happens in California, not in Phoenix, Arizona. In no one's memory had there ever been three severe flood years in a row.

Except for Murphy's Law we would be all right. But Murphy's Law did work. The greatest flood of all hit the Valley with all rivers carrying record flows. Every bridge was out except two. Damage to the properties of those whose business is in the rivers was greater than it had ever been. But the rock bed was intact. It would be tested in the spring and in the hot summer.

The Research and Development Phase was drawing to a delayed close.

The term "Research and Development" may give an image of white laboratory coats and work in sterile labs. As this particular project indicates, sometimes the work is in a hole in the ground in a driving rainstorm. Sometimes it is stop and go. Sometimes there is other work to do along with the project. Each project will be different. Each will be done on the basis of how much the promoter group is willing to "pay" to accomplish the fact. Some of the "pay" may be in the form of money invested. But the main price will be in time, disappointment, delay, hard work and being out of sync with original completion plans. If the project is important enough to finish, only one ingredient will suffice. That is perseverance.

GENERAL INFORMATION that relates to cooling and heating with rocks:

(1) A rock bed solar storage unit is a mass that operates, in a sense, like a storage battery. Consider degrees of heat stored as a positive charge and degrees of coolness stored as a negative charge. When the temperature of the "battery" is between 70°F and 75°F (the operating temperature of a building) the unit could be considered a "dead battery."

(2) The rocks can be given a positive charge (heated) from warm ambient daytime air or with air from a solar collector. A negative charge (cooled) can be made with cool night air, with night air cooled by a collector radiator using night sky radiation, or with an air conditioner.

(3) The amount of energy (heat) that the battery will release in the positive charge or the amount it will accept in the negative charge is stated in terms of British Thermal Units (BTU) or in terms of kilowatts (kw.).

The amount of energy required to raise the temperature of a pound of water 1°F is 1 BTU (3419 BTU = 1 kw.).

As a rule of thumb, it takes 0.20 BTU to raise the temperature of a pound of rock 1°F.

Therefore, this is a way of saying that the pound of water has stored 1 BTU and the pound of rock has stored 0.20 BTU; or that it will take five pounds of rocks to store as much energy as one pound of water with the same temperature rise.

(4) The following chart shows the storage capacity of various tonnages of rocks that have been given a charge of 10°F

differential from building operating temperature. Mathematical Example:

(10 tons rocks)

20,000 lbs x .20 = 4,000 BTU x 10 = 40,000 BTU

$\frac{40,000}{3419}$ = 12 kw.

SIZE	BTU at 10°F	KW EQUIV.
1 ton	4,000	1 kw.
10 ton	40,000	12 kw.
40 ton	160,000	47 kw.
60 ton	240,000	70 kw.
100 ton	400,000	117 kw.
200 ton	800,000	234 kw.
300 ton	1,200,000	352 kw

(5) *Insolation* refers to the solar radiation which arrives on earth, and its rate may be measured in terms of BTU per square foot per hour.

Insulation refers to material with a low rate of conductivity which will retard the passage of energy.

(6) The demand on a cooling or heating system can be determined by past energy use based on the temperature of the outside air. It may also be forecast based on the type of materials used to construct a building (estimated heat loss). The unit to which energy use is calculated is called a degree day.

A degree day is a unit that represents one degree of declination from a given point (65°F) in the mean daily outdoor temperature and is used to measure heat requirements.

(7) Energy use divided by Degree Days equals units of energy per Degree day. Degree days can be obtained from the Weather Bureau. The information is available on a year-round basis for both cooling degree days and heating degree days. This information is one of several approaches used in determining the proper size rock bin and proper size solar collector for a particular climate.

(8) Example —

Assume Heating Degree Days for one month at 1250.

113

Assume that it took 5,000 kw electricity to heat the building.

$$\frac{5,000 \text{ kw.}}{1,250 \text{ D.D.}} = 4 \text{ kw. per Degree Day}$$

If it is known that the degree days for a different period was 800, multiply 800 x 4 = 3,200 kw. which the building will use for that period. This system of calculation is fairly close to what the actual meter reading will be. (If the fuel used is oil or gas, use the units gallons or cubic feet instead of kw.)

(9) Although it takes five times the tonnage of rocks as it does water to store like amounts of energy, there are advantages, in many regions, in favor of rocks. Rocks do not freeze, leak or corrode. They lend themselves well to an air system. Usually, it is less costly to place rocks in a hole in the ground than to install a steel tank for water. This does not mean that rocks are better than water. Which storage medium is chosen will depend upon what is to be accomplished and on the economic facts which may make a positive difference between the two.

(10) There are many methods of using rocks for cooling and heating buildings. The one described here is only one of them designed for the Phoenix area. It is designed to both cool and heat. If it is to only heat, much smaller rock beds can be used. It is much easier and less costly to heat the rocks 30°F to 50°F over building operating temperature than it is to cool that many degrees below the building comfort zone. Therefore, a very large tonnage of rocks is required for the cooling load in the desert Southwest U.S.A. Low temperature differentials allows use of unsophisticated equipment.

(11) In general, the air voids will range from 30% to 45% in a rock bin. If the rock is fairly round and even in size the air voids will be greater than for rocks of mixed sizes. A rock bed with 1-1/8" to 12" rock had air voids of 38.8%. An even sample of 3" rock had air voids of 44%. The greater the air void in the bin, the less static pressure; thus, the less power required to move air.

(12) The larger the rocks are, the slower the discharge and recharge. 1-1/2 to 2-1/2 rock, either round or crushed, is a satis-

factory size.

(13) The farther the air must travel through the rocks, the greater the power requirements to run the fan. The coefficient of heat transfer from air to rocks or from rocks to air is high. Therefore, short air runs should be a goal.

(14) There are many ways to design a rock bin. The most inexpensive way is usually the best. In general, it is more economical to design the bin without structural walls or structural air dividers. The outside walls of the bin need only be designed to keep out water and rodents. The sloping sides of the bin are non-structural. The rocks just lay on the earth and on themselves. The cocoon to encase the rocks is a double thickness of 10 mil. polyethylene film. Other treatments can be added to this shape hole as required, such as pneumatically-placed concrete as a water barrier and/or insulation material.

CHAPTER XIV

THE PRODUCTION-MARKETING PHASE
—PROOF FOR A DECADE

The Production-Marketing Phase is a time of making and selling. Such a phase, if supported by many producers and marketers, can make a positive economic impact.

The fourth phase of a new business promotion is the Production-Marketing Phase. It is here that the entrepreneural group sees proof of their concepts and all of the work that was done before.

Most promotions for new industry would, upon attaining a goal, see a new factory building or an expansion to an existing operation. They would see product moving from the plant to market. They would realize the impact of success or failure rapidly.

In the case of the Rock Bed Heat Sink, we have only seen a structure develop that is to be produced by others. What we see is a construction scheme that is built with the present materials, the present tools, and the present technology of those now in the construction industry. Some parts and pieces of this energy plant may be bought from a local supplier or shop. However, the greatest part of the unit is produced on the building site.

There are few, if any, patents possible. No one has an exclusive right to build or to franchise. All classes of contractors can become involved. Many business-oriented persons or groups can promote and market the units. Against this amazing supply situation that is already in place is an equally amazing demand located wherever there is, or is to be, a building.

"So," you ask, "who can make any money promoting rock bed systems?"

It is obvious that the suppliers, the contractors and the marketing groups will have an income. "But, that is not like the neat, controlled income of a producing factory," you say. And I must answer that, "You are right!"

What must happen is more of a "movement" toward making buildings self-sufficient. It is more like promoting insulation and storm windows as energy savers. Many industries benefit from this promotion.

We are proposing the construction of a small energy factory to be connected to a building. Each building, so equipped, will be an income-producing entity. If the building is fully self-contained it will be producing more energy than it uses or, as we have alluded, be profitable.

In the selection of those to promote rock heat sinks, I could have picked various contractor groups, or the air conditioning industry, or the sheet metal industry, or any number of other suppliers. It was evident, however, that there is a ready-mix concrete plant and/or a sand and gravel plant in about every community. Their interests and their customer types are about the same everywhere. They are directly connected to construction. They can supply a part of the components — rocks and concrete. Therefore, this industry, with help from many others, can be the effective ambassador for rocks as a most exotic solar energy material.

The industry will not see great profits from this endeavor. Perhaps as volume grows it will be a nice piece of additional business for them. At this writing, other than for a few test units, there is no market at all.

Of much more interest to the suppliers, contractors, and the nation in general is an advantage of a subtle nature. That is the fact that each rock bed, however small, frees up fossil fuel in direct proportion to the amount of energy it supplies to a building.

Such a factor makes the freed-up portion of fuel available to heavy equipment, trucks and other wheels of industry. This is an area in which alternate energy may not be ready to replace fossil fuel for a few years.

To be able to realize total supply of oil from our own reserves will not only make us free from those who control foreign oil, but availability will keep the wheels of industry turning; thus, there will be production of wealth and jobs to keep our economy going.

Our continued dependence on foreign oil will only continue with a patchwork of international political decision which can only end in depression, war and wretchedness.

Therefore, I believe that those who supply rocks and concrete and their construction counterparts have a profit motive and a patriotic ideal as they promote solar rock heat sinks.

The Production-Marketing Phase is just beginning. Only those with foresight will become involved. Only those who become invoved can profit.

The decade of the '80's will be one of show and tell, show and tell, show and tell. Real success will be measured late in the decade as numbers have made an impact and refinement of the units have evolved. Such a time will be a moment of pride for the lowly rock and for those who believed. This simple development, when totaled with the impact of other types of econergy projects, can be the start of a new and prosperous era in world industrial history.

It is a very unusual, but very important promotion. Only those who build, or supply, or use, will profit in the early stages. However, as large numbers of units are installed, the entire nation profits from less imports of oil, less exports of dollars, and more production of wealth at home.

CHAPTER XV

ONE THING LEADS TO ANOTHER

The wheel was one of the greatest inventions. It sparked a chain reaction to new applications that has never stopped. Most simple Econergy projects, even a hole full of rocks, leads to that which is to be next.

As the rock bed development was underway in its early phases, and as the project progresses through its production-marketing phase, other related things became evident, or will become evident. Several "other" things should be discussed here to indicate how they evolve or work with a new idea. Often the related projects become the principal, at a future time, using the parent idea as its root or as one of its incidental parts.

On one occasion I was reflecting on some of the happenings as we built our first rock heat sink. One of the happenings was that we dug through the silt overburden at about nine feet depth. Exposed was an aggregate conglomerate much like that of the Salt River channel. We expected to find this condition so we just placed about one foot of dirt back on the bottom to make a smooth floor.

As I thought about this old aquifer under our rock bed, it made a connection with one of my cattle ranch projects in eastern Colorado.

The ranch was located 65 miles east of Pikes Peak on the

high arid plains. The blue grama grass sod makes a great country when it rains. However, during long dry periods it is a problem to manage grass pastures for cattle.

The decade of the 1950's was one of the driest periods.

South Rush Creek winds through the ranch, but it was dry. The east pasture was six square miles in size and the only water holes were near the extreme west fence. It required that I ride every day to check the cattle and the areas of available winter feed.

"Red" was my favorite horse. He knew better than I the job to be done. This day it was to duck his head and pace into a hard north wind. I also rode with my head down and my hat brim covering my eyes.

Near the top of the hill I felt Red's hind leg drop in a hole but he recovered quickly. I assumed that he had stepped into a badger hole.

With my head ducked low against the wind I could see his right hind leg as we moved up the hill and broke over the brow to the flat. I thought I must be imagining things. His leg was muddy, even though it had not rained or snowed for months.

We turned around and found the hole. It was wet. This seemed strange in such a dry country, especially since it was about 200 feet above the creek.

Digging around this area for several days disclosed that there was a thin, fine-sand seam about two inches thick — not enough for a well.

With the help of the Soil Conservation Service we came up with a method to recover the water from this thin sand structure.

We would build a "horizontal" well by digging a trench several hundred feet back into the hill exposing the thin sand seam. A perforated pipe laid in crushed rock along this seam had the same effect as if we had 200 feet of water sand thickness. The fact that the sand was tight (low porosity and permeability) was offset with the length of the perforated pipe.

The sand produced a flowing well that runs a two inch pipe full of water all the time. We built several of these wells.

As I thought about this I began to imagine another type of

cooling source for a valley that experiences 115°F temperatures all too often. For those who have endured the hot summers of the area it is hard to realize that just eight to ten feet underground the temperature is about 65°F the year around.

I could imagine the old barren aquifer that I saw exposed under the rock bin lying under a large portion of the valley floor. At one time the water table was near the surface in this water gravel. Intensive pumping for irrigation has lowered the water table to 800 or more feet in many areas. However, the voids that once held water are now filled with air — cool air. A method to recover this cooness began to form, using the horizontal water well idea from my ranching days as an example.

To this idea I applied the concepts that make the Rock Bed Heat Sink workable. It then became evident that here is a rock heat sink that is already in place with millions of tons of storage. All it needs is to be tapped, a process quite simple in areas of shallow overburden.

It requires about the same construction capability as is needed to install a leaching field for a septic system. Many contractors can expand their business into this econergy project. (See illustrations).

The importance is that it will replace millions of barrels of oil, even when applied to a relatively small area along the Salt River. There may be many areas where this could be used or the principle could be applied. I will let some of you imagine what to do in areas of abundant cool water and what to do with the water after use for air conditioning.

Then there is another way to use the low-grade energy that can be stored in heat sinks. Noted in the chapter on "New Industry Prospects" is a section on thermal gradients. It points out that ammonia will turn to gas in the warm surface water (about 81°F) of a tropical ocean, and condense to liquid if exposed to the temperatures of the cool, deep water (about 43°F at 2,100 feet). Much has been written on this research on Ocean Thermal Energy Conversion (OTEC), and the several prototypes on barges floating around the warmer oceans making electricity.

The fact that these prototypes work and produce electricity, as expected, has not triggered a commercial application. The process is not complex. Most certainly, it cannot qualify as "space age" technology.

Herein, (see illustrations) is a schematic drawing of a system using freon as the expandable to turn a turbine. All of the components of this system can be purchased "off the shelf" and are now being produced. (Tanks, pumps, flywheels, heat exchangers, condensers, turbines, generators.) It is not necessary to go into the ocean to make it work. Water or rock heat sinks on land can maintain 50°F to 100°F temperature differentials if freon is designed to gas at about 70°F.

The system works on about the same principle as a refrigerator. From a tank, the freon is piped to a heat exchanger in a heat sink which is charged from a solar collector. In the exchanger the freon changes to a gas. As a gas it has expanded and moves under pressure to a turbine which turns an electric generator as the gas moves through the turbine to a condenser. From here it is pumped back to storage as a liquid to start the process over again.

The use of low-grade energy that can be obtained from low temperature differentials and low temperatures does not have the high energy impact to be obtained by super heated steam. However, one must remember that to make just one building electrically self contained does not require more than perhaps 100 kw per day maximum.

To make an analogy under which this low energy unit will work one must think back to the days when most farms and ranches had a one cylinder, four cycle gas engine to grind grain and do other small chores. Such an engine is designed to fire after the compression stroke and the energy of this combustion must carry the engine through its next three cycles. ((1) Power stroke, (2) exhaust stroke, (3) intake stroke, (4) compression stroke.)

The interesting thing about this engine is the fact that it will not run until one more thing is added. That is a flywheel. Without the flywheel the usable power generated by the burning gasoline is dissipated. With a heavy spinning wheel the energy is stored in motion (kinetic) and is then usable to carry the engine through the three "get ready" cycles and with energy left over to do other work.

The same would be true of any system that would use low-grade energy to make electricity. Heat at 120°F to 160°F can do a lot of work if the energy is stored in a flywheel.

Such a system will not have high efficiency in the use of total stored kw in the heat sink. For example, a rock bin of 200 tons will store 234 kw for each 10°F of temperature differential or 1170 kw at 50° differential. If the system is 10% efficient at 50°F differential, the building would have 117 kw per day to use. With a few innovations this percent of efficiency may be obtained or increased. This is sufficient.

Therefore, in the freon turbine electric generator is a type of project that can be manufactured in a factory. But it is more of an assembly of parts that can be purchased from others. Properly funded, about any community could have an attractive new industry of this type. The market is so big that it is difficult to estimate how many plants could prosper with this item.

As I look back on the activities of the past four years and the path that suggests that "One Thing Leads to Another," I am excited when I anticipate what future econergists will present. I can only guess at the power of thousands of people with their minds and actions aimed at winning the NEW INDUSTRIAL REVOLUTION.

Of greater interest is to know that what we believed in the decade of the forties as early "chemergists" is still true in the decade of the eighties. What I learned from some great men is now badly needed. What they passed on to me was worth refining and passing on to others.

Whoever dreamed, as we discussed the coming fossil fuel crisis thirty years ago, that all of the technology would be available to solve the problem today? And who could have believed that the only missing ingredient to the solution would be that the small business entity would not understand its role as the one to salvage the private economic system from disaster? Who could have imagined the power of the negative to hold back development in what had always been a positive American business institution?

However small the start, a few will be active econergists. Those few will convince many. From the many there will be new Econergy Industries and Projects developing everywhere.

Econergy is people who believe in doing sound business things to solve their nations' problems. Econergy is not a thing to be picked up in your hand and taken anyplace. It is a thing of understanding. It is a *movement*.

CHAPTER XVI

THE BEGINNING — NOT THE END

Everything has a beginning and an ending. The fossil fuel economy, which produced the greatest wealth ever known, is ending. The new economic order will be known as Econergy. It is just beginning. Like most things that start at the grassroot level, it must be sponsored by those who consider themselves "Just Folks." Everything worthwhile in Econergy lies ahead.

Most books arrive at the last chapter with a complete work. The author is proud to write the final phrase—"The End."

ECONERGY — THE CONCEPT OF PLENTY has no end. It will always be charged with new beginnings as long as the world expands into the use of renewable resources.

The basic concept has been evident for many years. It has not changed. The change has been in the social, political and economic condition of the world.

It is in what must be done and what can be done that we find the greatest of the new beginnings. Never before have ordinary men and women been called upon to solve a world crisis with a movement toward a more local or regional business form.

Early on it was pointed out that to solve the national or world problem of food, clothing, shelter and energy is almost impossible. However, your problem and my problem is small. Our community problem is small. We can identify this size problem.

125

We can do something about it through applications of the econergy concept.

This basic, generalized primer describes the concept of utilizing the renewables that was given this writer following World War II. It is now passed on to those of you who care enough to think and to act. To you is entrusted the future of econergy. To you is entrusted the refinements in the concept which must follow.

It is understood that by asking only those who care enough to *think* and *act* to become the *real* Econergists will eliminate about 90 percent of the population. But the remaining 10 percent of the population is a group which constitutes millions of people; and millions of people thinking and acting toward the goals of econergy will win the next industrial revolution.

Locked in the minds of the so called common man or woman is a wealth of knowledge that has never before been tapped. There has never been a place where this knowledge could be *exposed* and, in effect, *harvested*.

Again, some are saying, "Who, me?" And again, "Who else but you?"

Intelligence is not found only in high officials of government and large business institutions. The common person, the laborer, the housewife, the clerk, the secretary; as a group, are a reservoir of intelligence far greater than at the executive level because there are more of them. All have inherent knowledge because all are products, both physically and mentally, of their thousands of ancestors. An ancient ancestor may have passed a rare chromosome to you that gives you understanding and knowledge not known by any other.

A place for you exists in the econergy movement. A place is described here into which you can have an input. That place is one that you can generate right where you live now.

So again you ask, "What can I do in this complex world of bigness?"

We see a world of giantism into which we find it difficult to interject our ideas. It seems that this giantism covers all things as if it were an umbrella under which all things must function.

The financial community, the industrial community, the educational system and many other social and economic complexes make up the tightly woven fabric of this so-called umbrella.

At times, the fabric of this cover shields all and things are safe and tranquil. Most of the time the cover is full of holes and this leads to frustration in politics, business and individuals that must function under the umbrella.

However, this umbrella must exist with its complex fabric. It represents a thing called, "The way it is," and, with that said, the term is defined.

Those of us who are Econergists have the time and a way to change the fabric in this thing called "The way it is." We can weave the new threads of renewable resources into local and regional production. We can add thread after thread, no matter how small, year after year for generations:

We can slip under the protection of the umbrella and do our many small things. Most of the world will not notice, few will believe it is possible. By so doing, we will quietly change from the dying fossil fuel economy which supports the present, unyielding giantism to Econergy — The Concept of Plenty.

The age of renewable resources will, therefore, *evolve* rather than be *forced* on the present system through new restrictive laws.

The key to the problem lies in the answer to your question, "What can I do?"

Realistically, each of you may not go out into the world and promote a new company. It is hoped that your understanding of a promotion was enhanced as you read about "Promotion of New Companies." You saw a different approach in the example on the promotion of the Rock Bed Heat Sink.

In any event, it is evident that most things happen as a team effort. Clubs, associations, committees and groups do not often win precise goals. But, a team, guided by a leader, can play a game to win.

If it becomes possible to have millions of active Econergists, it is evident that they must act in small groups; and the best group form is a *TEAM*.

However, before we discuss the ECONERGY TEAM, we should consider the individual team member — YOU and I.

Your first thoughts may be that you already have too much to do. You work and have other outside interests. You are busy. You do more than your share now. Why get involved in this new thing?

The answer to all of these may lie in the answer to another question, "What are you and I willing to do to preserve and enhance our FREEDOM?"

There is no need to scoff at such a lofty question. You and I are not really free. We are captives of a few major money pools and a few oil cartels.

These entities control our lives through their positive control of fossil fuel. There is no need to dispute this. We have all witnessed, in the past decade, the world's most powerful nation being forced to make domestic and foreign policy from a stance of appeasement to this powerful group. This nation is a captive. As a result the citizens of this country are no longer free. The predicament limits our personal freedom in social and economic choices. The situation will gradually worsen as fossil fuel reserves dwindle.

If our nation is forced to cower under the threat of withheld oil supply, where does this place you and me in the scheme of things?

The answer lies in what we can do and must do from our grassroots habitat. If these things are very simple and perhaps enjoyable, are we likely to get them done?

Let us look at what an ECONERGY TEAM can do and how they can do it.

First is the premise that there is a strong desire among many to be self-sufficient and free. This group is looking for a positive approach to the many problems that confront the society of today. However, they turn away from the frustrated responses to these problems by those groups who choose the negative "March in the street protest" type of civil disobedience.

128

Those who would participate as a leader or member of an ECONERGY TEAM can make an impact from a most unusual starting activity. It is initiated by exercising one of the basic arts from the law of giving and receiving. It is often called the art of conversation and, in recent years, has been referred to as the "lost art."

Could it be that a structured effort in talking, listening, thinking and understanding, by those who consider themselves average citizens, will make an impact toward freedom and self-sufficiency? Could this be a type of "Think Tank" operation? What else is a think tank than a place where ideas are exchanged? It is a place of structured conversation. It is called communication. It is called visiting. Yes, an impact can be made by the so-called, average person. Yes, it can be called a think tank. It can also be called an ECONERGY TEAM.

A team should consist of eight to sixteen members. They should meet at least once a month. They should discuss and understand as much as possible about the various aspects of econergy as this book indicates. They should add to the concept. They should pick a project of econergy that fits their area and their interests. They should continue to meet even though they do not easily find a project on which they can act. They should have a different guest each session. They should encourage the formation of other teams. They should exchange ideas and expertise with other teams. They should enjoy the brainstorming activity. Each member will listen and each will participate in the discussion (give and receive).

This activity can be started by one who will be a leader by the simple act of inviting other couples or individuals to join the team. Also, one who may not wish to lead can initiate a team by asking another to be a team leader with whom he or she can participate.

As the number of teams increase, it is obvious that formal organizations of teams will develop — first in groups of teams in local areas and then to state and national organizations.

Since anyone can form a team there will be diversity of capability and interest in promotion, finance, banking, politics, law, management, marketing, technology, innovation, communication and econergy team organization.

The interplay of team diversities will require a strong national organization through which information can be easily and quickly exchanged. All of this can start from the enjoyable activity of your monthly or weekly meeting with your friends and team members.

There will be monthly tapes, video tapes and newsletters to spark the topic of the "Think Tank" sessions.

There will be activities which will produce income for ECONERGY TEAMS. There will be no dues for individual team members. The team is the entity through which all will act and react to state and federal team organizations.

The details of the initiation of this movement will be explored and exposed in other brochures and booklets. However, the real movement and organization will be guided and developed by those teams who are first to become active.

Activities of a team should grow through the stages of: 1) understanding, 2) support for, 3) communicator of, 4) aid to, 5) initiator of, 6) promoter of, 7) developer of, 8) consumer of. Hopefully, some teams will have experience and expertise to do it all from the date of their organization. However, this sort of group may be the exception.

1) *UNDERSTANDING* is most important if energy independence and personal freedom is to be initiated from the grassroot level. Following understanding all other activities fall into place.

2) *SUPPORT FOR* those who produce from the renewable storehouse is a great need. It is most helpful when it comes from the local area of production. It may be evidenced by a few kind words, or it may require active and decisive action. This is probably the most logical first activity for a team following its effort to become informed.

3) *COMMUNICATOR OF* ideas transfers understanding to a

community in general or perhaps to a business entity in particular. This is an activity of *merit* and is a level of econergy to be reached by all teams.

4) *AID TO* a particular project or business is that of actually giving service to promoters of new business expansions. This is basic participation and is a degree of *high merit*.

5) *INITIATOR OF* a new business expansion that will be promoted by others is a feature of innovation and a part of the creation phase of promotion. This would be a degree of achievement of *superior merit* for any econergy team.

6) *PROMOTER OF* new econergy projects will be a level of action, the merit of which will be in the degree of *excellence*. Those now doing this, and there are some, will hold a place of affection and honor in the econergy movement.

7) *DEVELOPER OF* econergy projects by a team or by others has reached a degree of the *elite in merit*. These are the people who really do it. They are the ones to which all teams strive to help to the limit of each team's commitment and capability.

8) *CONSUMER OF* devices, equipment, systems and programs which use the econergy concept is a degree of achievement to which all econergists can function. There may be no higher degree of merit than to use the renewable resource to which may be given the status — *merit of experience*.

The econergy team activity will be one of your most rewarding experiences. You will be participating in that which is to be next, and you will be a part of an answer to a world problem for which there are few solutions proposed. There are none, except the econergy program, in which you have been asked to participate or to even hear about. The reason is that there are no other workable solutions being offered. Your participation in this project is paramount.

How strange this approach must seem as it is first considered. But again, what other plans are presented?

Rest assured that it may not be easy. It was not easy a generation ago. Many in government and industry opposed the use of renewables in the decades of the 1930's and 1940's. There are interests that fear this new industrial order. Many wish to keep the status quo.

Today's great need caused by the dying fossil fuel economy will not make it easy now. There are vested interests. There are negative, doubting people. They resist the greatest new ideas.

But who will notice a few econergy teams? Who will believe that a great new order is being born? Who, except us?

There is no need to be against our major business institutions. Do not fight "the way it is." We are a part of it. We need them and soon they will need us.

There is only need to slowly change "the way it is" by allowing the persistent growth of econergy to evolve into a new economic order.

No need for the big hyperactive program. No need for major funds. No need for government interference. No room for the negative.

This, again, brings us back to just you and me. What we have done is not important. What we have learned to do next is.

This is the beginning.

CHAPTER XVII

ILLUSTRATIONS

**IDEA PICTURES TO IGNITE
YOUR ENTREPRENEURALSHIP**

ILLUSTRATION I

THE ROCK BED HEAT SINK **STORAGE IN MASS**

THIS IS JUST A HOLE IN THE GROUND FULL OF ROCKS. ITS MAIN FUNCTION IS TO HEAT OR COOL BUILDINGS. IT OPERATES AS A STORAGE BATTERY.

HEATING MODE RECHARGE FROM:
- COLLECTOR
- WASTE HEAT
- AMBIENT AIR (DAY)
- SUPPLEMENTARY HEATER

COOLING MODE RECHARGE FROM:
- AMBIENT AIR (NIGHT)
- COLLECTOR (NIGHT SKY RADIATION)
- SUPPLEMENTARY COOLER

THIS DESIGN MAY BE SATISFACTORY FOR LARGER UNITS TO OPERATE THE FREON ELECTRIC GENERATOR. (SEE ILLUSTRATION).

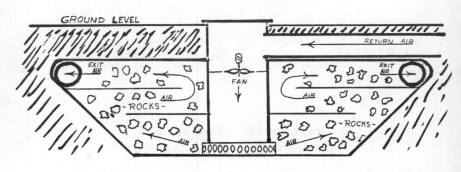

ROCK BED   SIDE VIEW

ILLUSTRATION II
ROCK BIN (TOP VIEW)

THIS VIEW IS OF THE TOP LAYER OF ROCKS IN THE BIN
SHOWN IN THE PREVIOUS ILLUSTRATION (SIDE VIEW).
THE PERIMETER PIPE EQUALIZES AIR FLOW' THUS' BIN
TEMPERATURE IS EQUALIZED.

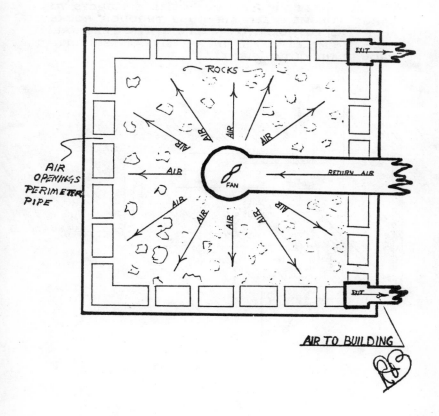

ILLUSTRATION III
A SIMPLE ROCK BIN WITHOUT STRUCTURAL WALLS OR SUPPORTS

ENERGY TRANSPORT BY AIR ------ ENERGY STORAGE BY ROCKS

CONSTRUCTION TRADES TO BUILD:
      ENGINEERING (EXCAVATION & ROCK PLACEMENT)
      SHEET METAL (DUCT WORK)
      ELECTRICIAN (FAN MOTOR & CONTROLS)

USE ADEQUATE AIR VOLUME --- SIZE AIR DUCTS TO LARGE VOLUME. KEEP AIR RUNS THROUGH ROCKS SHORT AS POSSIBLE TO REDUCE STATIC PRESSURE. KEEP IT SIMPLE. KEEP IT INEXPENSIVE. IT IS UNDERGROUND AND OUT OF SIGHT.

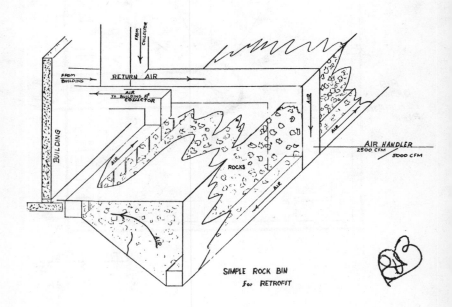

SIMPLE ROCK BIN
for RETROFIT

ILLUSTRATION IV
SCHEMATIC ROCK BED HEAT SINK
FOR HEATING AND COOLING BUILDINGS

DAMPER MODES:	OPEN	CLOSE
AIR TO BUILDING	6, 7	
TO RECHARGE ROCKS:		
FROM COLLECTOR	1,3,5	2,4
	(RECHARGE FOR HEAT BY DAY)	
FROM AMBIENT AIR	1,2,4,5	3
	(RECHARGE FOR COOL AT NIGHT)	

FAN ON DELTA-T CONTROL & THERMOSTAT
CONTROLS - AUTOMATED (SOLID STATE)

A MULTILEVEL RECHARGE SYSTEM
OPERATED BY ONE FAN

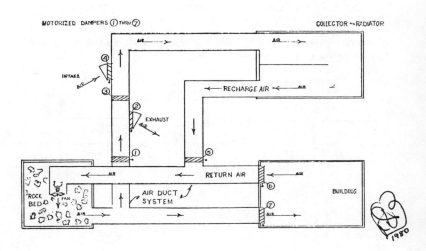

ILLUSTRATION V
COOLING FROM A BARREN AQUIFER

LOCATION EXAMPLE:
 PHOENIX, ARIZONA --- THE SALT RIVER VALLEY THERE ARE THOUSANDS OF ACRES OF HOMES AND WAREHOUSES THAT COULD TAP THIS COOLING SOURCE.

COOLING CAPACITY FORMULA:
 A LOT SIZE 200 FT X 200 FT WITH 30 FT OF GRAVEL DEPTH WOULD HAVE 800,000 CU FT OF HEAT SINK IN PLACE. THIS WOULD STORE .20 BTU PER LB PER 1°F - ROCK TEMP. 65°F - BLDG. TEMP. 75°F - DIFFERENTIAL 10°F. (ASSUME WT OF ROCK 110 LB PER CU FT)
  800,000 X 110 X .20 X 10 $=$ 176,000,000 BTU
  176,000,000 $\div$ 3419 $=$ 51,477 KW
  51,477 KW @ $.06/KW $=$ \$3,089.00 OF COOLING ALREADY IN PLACE TO BE USED FOR THE COST OF RUNNING TWO FANS.

NOTE: COOL WATER CAN BE USED THE SAME WAY BY PUMPING IT TO A HEAT EXCHANGER.

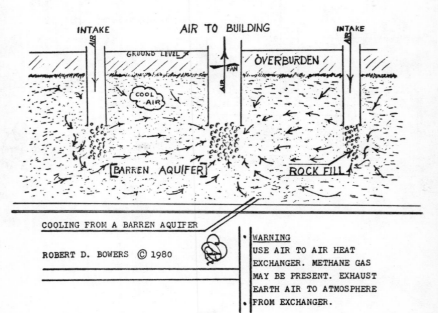

COOLING FROM A BARREN AQUIFER

ROBERT D. BOWERS © 1980

WARNING
USE AIR TO AIR HEAT
EXCHANGER. METHANE GAS
MAY BE PRESENT. EXHAUST
EARTH AIR TO ATMOSPHERE
FROM EXCHANGER.

ILLUSTRATION VI

SOLAR ELECTRIC GENERATOR

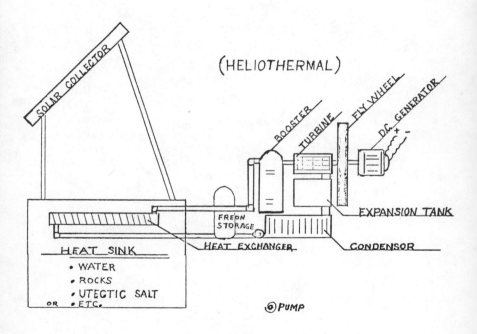

(HELIOTHERMAL)

SOLAR COLLECTOR

BOOSTER

TURBINE

FLY WHEEL

DC GENERATOR
+ −

EXPANSION TANK

FREON STORAGE

HEAT EXCHANGER

CONDENSOR

HEAT SINK
- WATER
- ROCKS
- UTECTIC SALT
OR • ETC.

⊚ PUMP